The Proprotein Convertases: Discovery, Characteristics, and Link to Tumor Progression and Metastasis

Colloquium
Digital Library of Life Sciences

This e-book is a copyrighted work in the Colloquium Digital Library—an innovative collection of time saving references and tools for researchers and students who want to quickly get up to speed in a new area or fundamental biomedical/life sciences topic. Each PDF e-book in the collection is an in-depth overview of a fast-moving or fundamental area of research, authored by a prominent contributor to the field. We call these e-books *Lectures* because they are intended for a broad, diverse audience of life scientists, in the spirit of a plenary lecture delivered by a keynote speaker or visiting professor. Individual e-books are published as contributions to a particular thematic **series**, each covering a different subject area and managed by its own prestigious editor, who oversees topic and author selection as well as scientific review. Readers are invited to see highlights of fields other than their own, keep up with advances in various disciplines, and refresh their understanding of core concepts in cell & molecular biology.

For the full list of published and forthcoming Lectures, please visit the Colloquium homepage: www.morganclaypool.com/page/lifesci

Access to Colloquium Digital Library is available by institutional license. Please e-mail info@morganclaypool.com for more information.

Morgan & Claypool Life Sciences is a signatory to the STM Permission Guidelines. All figures used with permission.

Colloquium Series on
Protein Activation and Cancer

Majid Khatib, *Ph.D., Research Director, INSERM,*
and University of Bordeaux, France

This series is designed to summarize all aspects of protein maturation by convertases in cancer. Topics included deal with the importance of these processes in the acquisition of malignant phenotypes by tumor cells, induction of tumor growth, and metastasis. This series also provides the latest knowledge on the clinical significance of convertase expression and activity, and the maturation of their protein substrates in various cancers. The potential use of their inhibition as a therapeutic approach is also explored.

For a full list of published and forthcoming titles:
http://www.morganclaypool.com/page/pac/1/1

The Proprotein Convertases: Discovery, Characteristics, and Link to Tumor Progression and Metastasis
Abdel-Majid Khatib
www.morganclaypool.com

ISBN: 9781615045365   paperback

ISBN: 9781615045372   ebook

DOI: 10.4199/C00072ED1V01Y201301PAC005

A Publication in the

COLLOQUIUM SERIES ON PROTEIN ACTIVATION AND CANCER

Lecture #5

Series Editor: Majid Khatib, Ph.D., Research Director, INSERM, and University of Bordeaux, France

Series ISSN
ISSN 2169-9399   print
ISSN 2169-9410   online

The Proprotein Convertases: Discovery, Characteristics, and Link to Tumor Progression and Metastasis

Abdel-Majid Khatib
University of Bordeaux, France

COLLOQUIUM SERIES ON PROTEIN ACTIVATION AND CANCER #5

ABSTRACT

Proprotein convertases (PCs) are a family of proteases including PC1, PC2, Furin, PC4, PACE4, PC5, and PC7. These enzymes are involved in the maturation of many precursor proteins involved in the process of tumorigenesis and metastasis. Since their discovery, PCs were suggested as potential targets for anti-cancer therapy, and their activity was found to directly affect tumor cell proliferation, migration invasion, and the malignant phenotypes of tumor cells. Here, we discuss a number of previous and recent findings on the PCs features, their implication in the regulation of multiple cellular functions that impact on the invasive/metastatic potential of cancer cells, and their clinical relevance in cancer patients.

Among the substrates of the proprotein convertases, various growth factors, their receptors, adhesion molecules, and proteases were identified. The PCs are inhibited by endogenous and exogenous inhibitors. To date, only pro7B2, a specific chaperone of PC2, and the granine-like precursor of neuroendocrine protein proSAAS, a selective ligand of PC1, have been identified as endogenous inhibitors of the PCs found in the regulated pathway. However, only PCs prosegments, several bioengineered inhibitors, peptides, and non-peptide compounds were found to inhibit the activity of the PCs found in the secretory pathway.

KEYWORDS

proprotein convertases, furin, PACE4, PC5, PC7, inhibitors, cancer, metastasis, clinical relevance of PCs

Contents

Abbreviations

ER	endoplasmic reticulum
TGN	trans-Golgi Network
PCs	proprotein convertases
PACE	paired basic amino acid cleaving enzyme
SPC1	subtilisin-like proprotein convertase
SKI-1	subtilisin/kexin-like isozyme-1
Narc-1	neural apoptosis-regulated convertase 1
TIMP	tissue inhibitor of metalloproteinase
CRD	cys-rich domain
HSPGs	heparan sulphate proteoglycans
EC	endothelial cells
IC	inflammatory cells
ECM	extracellular matrix
MT-MMPs	membrane-type matrix metalloproteinases
ICAM	intercellular adhesion molecules
VCAM	vascular cell adhesion molecule
MadCAM	mucosal addressin cell adhesion molecule
FAK	focal adhesion kinase

CHAPTER 1

Discovery of the
Proprotein Convertases

Proteolysis is a cellular process responsible for the conversion of various protein precursors with high molecular weight peptides as well as small polypeptides. This process constitutes the main post-translational modifications that occur during the intracellular transport of proteins from the endoplasmic reticulum (ER) to the trans-Golgi network (TGN), in the secretory granules (SG) or on the surface of various cells. In general, during proteolysis, the protein precursors are produced in inactive forms and activated following their cleavage (growth factors and proteases) [1–2]. However, some active unprocessed proteins seemed to lose or reduce their activity after cleavage (FGF-23, Sema 3B) [3], while for other precursors, the cleavage does not appear to affect their functions (integrin $\alpha4$–$\beta1$). There are also biologically active protein precursors whose cleaved forms induce an opposing function (NGF) [4–5] (Figure 1). Among the enzymes responsible for these processes, the proprotein convertases (PCs) occupy a primordial place.

The history of proprotein convertases began in 1967, when the group of Steiner has demonstrated for the first time that insulin was a byproduct of post-translational processing of its precursor, proinsulin [6]. At the same time, Chrétien M and Li CH, by analyzing the sequence of the POMC (proopiomelanocortin), found that the sequence of MSH (melanocyte-β-stimulating hormoneβ lipotrophin hormone) contains the β-LPH (β *lipotrophin hormone*) sequence suggesting the existence of a proteolytic mechanism responsible for the maturation of the precursor protein POMC to small peptides [7]. Only 20 years later, the enzymes responsible for these processes were identified and called proprotein convertases (PCs). These serine proteases were found to be calcium-dependent enzymes and are related in their structure to bacterial subtilisin and to yeast kexin. Indeed, initially, the discovery of the proprotein convertases began when the Kex2 has been identified in yeast as a subtilisin-like endoprotease responsible for the cleavage of certain protein precursors [8]. Subsequently, in 1989, the analysis of a database identified a structure similar to the Kex2 [9]. This structure is part of a product of the fur gene (*c-FES/FPS* upstream region) previously identified by the group of Roebroek et al. and was so named due to its proximity to the

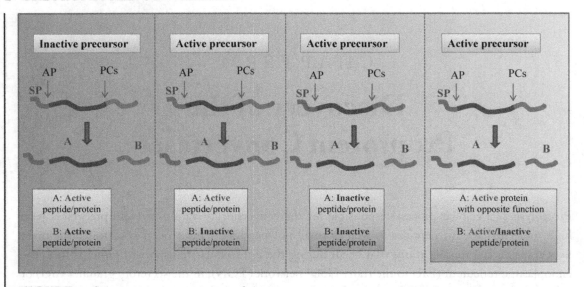

FIGURE 1: Schematic representation of the maturation of active and inactive precursor proteins by the pro-protein convertases. The cleavage of substrates generates active and/or inactive protein/peptide (A and B). SP: Signal peptide, AP: Aminopeptidase site and PCs: Proteolytic cleavage site recognized by the pro-protein convertases.

proto-oncogene *c-FES/FPS* [10–11]. The product of this gene, Furin, also called paired basic amino acid cleaving enzyme (PACE), or subtilisin-like proprotein convertase 1 (SPC1) was the first member of the family of human PCs to be identified [12]. To date, Furin is also known as PCSK3. Lately, the use of PCR in the presence of degenerate oligonucleotides based on the sequence of the catalytic domain of Kex2 allowed the identification of other PCs named: PC1/PC3/PCSK1 [13–14], PC2/PCSK2 [13, 15], PACE4/PCSK4 [16], PC4/PCSK5 [17–18], PC5/6/PCSK6 (isoforms A and B) [19–21] and PC7/LPC/PC8/SPC7/PCSK7 [22–24]. Lately, two other members of the family of PCs were identified: namely, SKI-1/S1P/PCSK8 (subtilisin/kexin-like isozyme-1) [25–26] and Narc-1/PCSK9 (neural apoptosis-regulated convertase 1) [27–28].

Based on the sequence of the cleavage site recognized by the PCs, these enzymes were listed in three different groups: the first group includes the Kexin-like PCs that cleave precursor proteins at the basic site (K/R)–Xn–(K/R), where X can be any amino acid except Cys, and $n = 0, 2, 4$ or 6. The second group contains currently only the convertase SKI-1/S1P/PCSK8 that cleaves substrates within the cleavage site (K/R)-X-(hydrophobic)–(L, T). The last group contains only PCSK9/Narc-1 and seems to have a preference for the VFAQ sequence as deduced from its autocatalytic cleavage site and has no substrate known to date.

1.1   STRUCTURE OF THE PROPROTEIN CONVERTASES

The PCs present a high structural homology. They all have an N-terminal structure with common functional areas and variable C-terminal domain structure (Figure 2).

The N-terminal domain of the PCs is the most conserved structure. This domain contains primarily a signal peptide, which is excised early in the ER, a process necessary for the entry of the enzyme in the secretory pathway, followed by the prosegment that acts as an intramolecular chaperone protein essential for the protein folding and behavior as an inhibitor of the cognate

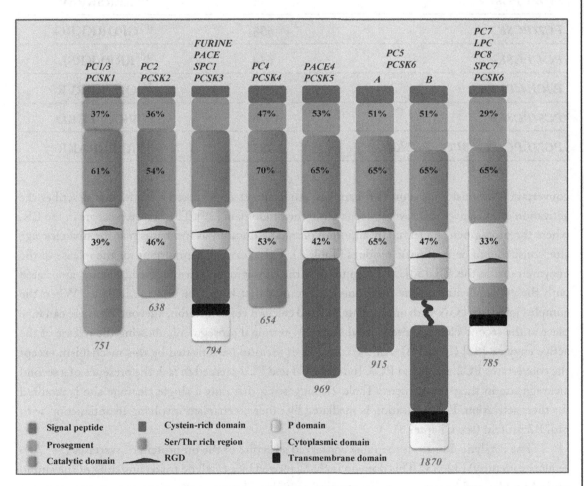

FIGURE 2: Schematic representation of the structure of the proprotein convertases. The numbers indicate the homology percentage of each area as compared to that in Furin structure. Some convertases have many names because they were discovered at the same time by different groups. The number of amino acids for each convertase is also indicated.

TABLE 1: Amino acid sequence of the autocatalytic site of PCs. For each PC, the number of amino acids and the sequence of its autocatalytic site are represented.

PROPROTEIN CONVERTASES	AMINO ACID NUMBER	AUTOCATALYTIC SITES
Furin/PACE/SPC1/PCSK3	794	[101]AKRRTKRD
PC1/3/PCSK1	751	[105]KERSKRSV
PC2/PCSK2	638	[103]GFDRKKRG
PC4/PCSK4	654	[105]RRRVKRSL
PACE4/PCSK5	969	[141]QEVKRRVK
PC5/PCSK6	1870	[109]VKKRTKRD
PC7/LPC/PC8/SPC7/PCSK6	785	[134]RLLRRAKR

convertase. Previously, based on Furin studies, Anderson et al. proposed a model that describes the activation of PCs involving several steps within the prodomain [29]. The first step occurs in the ER, where the prosegment is cleaved by an intramolecular autocatalysis mechanism at the first cleavage site, consisting of pairs of basic residues (Table 1). This cleavage is important for the release of the enzyme outside the ER [30]. We assume that the prosegment, once cleaved, remains associated with the enzyme catalytic site by forming a complex that keeps the enzyme inactive. When the complex joins the TGN with appropriate pH and calcium concentration, a second cleavage can take place at the second cleavage site located on the N-terminal prosegment, allowing the release of the active enzyme [31] (Figure 3). All the convertases seem to be activated by this mechanism except the convertases PC2, PC4, and PC7. Indeed, PC4 and PC7 seemed to lack the presence of a second cleavage site in their prosegment (Table 2), suggesting that only a single cleavage site is required for their activation. PC2 activation is mediated by other mechanism involving its interaction with pro7B2 protein (see Chapter 3).

The catalytic domain is the most conserved structure of the proprotein convertases (54–70% sequence identity) [32–33]. This domain is characterized by a catalytic triad consisting of the amino acids Asp, His, and Ser necessary for enzymatic activity. The active site of the enzyme also contains an Asn residue that ensures the stabilization of the pocket of the oxyanion transiently formed during the hydrolysis reaction. The position of these residues is conserved except for PC2, where the Asp residue replaces the Asn residue. Further studies revealed that the Asp residue might participate in the activation of PC2 in the absence of an optimum pH [34].

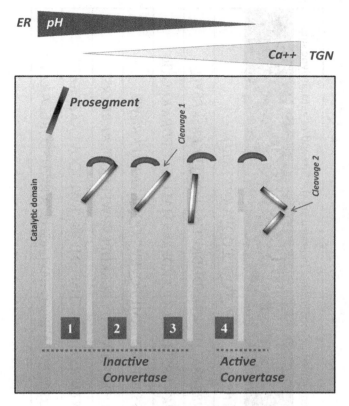

FIGURE 3: Schematic representation of the mechanism of PC inhibition by their prosegments. In the endoplasmic reticulum (ER), inactive convertase is subject to folding events and peptide signal removal **(1)**, in the late secretory pathway, PC undergoes a first autoproteolytic cleavage of its prosegment **(2)** that remains associated with its inactive enzyme **(3)**. At low pH and higher [Ca2+], the complex is dissociated leading to PC activation **(4)**.

The P domain, also called HomoB, or "middle domain," is located on the C-terminal of the catalytic region (Figure 2). This domain is essential for the proper folding of the enzyme, its stability, and activity. This domain also allows for the maintenance of an adequate concentration of Ca+ and pH required for the activity of the PCs [35]. This area also contains an RGD motif conserved in all convertases except for PC7. The role of this motif is not yet well known. However, mutation of this motif in PC1/3 resulted in a loss of function followed by a relocation of the enzyme: instead of going through the regulated secretory pathway, the enzyme progressed toward the constitutive secretory pathway [36].

The C-terminal domain is more variable and specific to each proprotein convertase. Only the C-terminal domain of the convertases Furin, PC5/6, and PACE4 has a Cys-rich domain (CRD).

TABLE 2: The amino acid sequence of the proprotein convertase prodomains. Indicated are the first and/or autocatalytic sites.

PC PROSEGMENTS	PEPTIDES SEQUENCE AND THEIR CLEAVAGE SITES
ppFurin	YHFWHRGVTKRSLSPH*RPRHSRL*QREPQVQWLEQQVA*KRRTKR*DVY
ppPC1	IGSLENHYLFKHKSHP*RRSRR*SALHITKRLSDDDRVTWAEQQYE*KERSKR*SVQ
ppPC2	PFAEGLYHFYHNGLA*KAKRRR*SLHHKQQLERDPRVKMALQQEGF*DRKKKR*GYR
ppPACE4	EDYYHFYHS*KTFKR*STLSSRGPHTFLRMDPQVKWLQQQEV*KRRVKR*QYR
ppPC4	IFPDNQYFHLRHRGVAQQSLTPHWGHPLRLKKDPKVRWFEQQTL*RRRVKR*SLV
ppPC5	NIGQIGALKDYYHFYHS*RTIKR*SVISSRGTHSFISMEPKVEWIQQQVV*KKRTKR*DYD
ppPC7	RIGELQGHYLFVQPAGHRPALEVEAIRQQVEAVLAGHEAVRWHSEQRLL*RRRAKR*SVH

PROPROTEIN CONVERTASES	MOUSE		HUMAN	
	CHROMOSOME	LOCI	CHROMOSOME	LOCI
Furin/PACE/SPC1/ PCSK3	7	7D1-E2	15	15q25-q26
PC1/3/PCSK1	13	13C2	5	5q15-q21
PC2/PCSK2	2	2G1	20	20p11.2
PC4/PCSK4	10	10C1	19	19p13.3
PACE4/PCSK5	7	7B5	15	15q26
PC5/PCSK6	19	19B	9	9q21.3
PC7/LPC/PC8/ SPC7/PCSK6	9	9A5.2	11	11q23.3

TABLE 3: Chromosomal location of human and murine PC genes

This domain allows PCs when secreted to remain bound to the cell surface and interact with the tissue inhibitor of metalloproteinase-2 (TIMP-2) through the heparan sulphate proteoglycans (HSPGs) of the extracellular matrix [37–38]. Indeed, CRD domain deletion or cell treatment with heparin was found to prevent the binding of these convertases to the cell surface [38].

The Furin, PC7, and PC5B are the only PCs that present at their C-terminal region, a transmembrane domain. This area allows these PCs to move and cleave substrates between the TGN and the cell surface. Conservation of the catalytic domain and the P domain in all PCs suggests that the PC genes are derived from a common ancestral gene [4]. The study of the chromosomal localization of PCs revealed that only Furin and PACE4 seem to share the same chromosome (Table 3).

1.2 CELLULAR LOCALIZATION OF PROPROTEIN CONVERTASES

Various studies on the expression of PCs in different tissues allowed the classification of PC distribution into four groups [1] (Table 4). The first group includes Furin, PC7, and PC5B. The convertases Furin and PC7 are expressed in most tissues and cell lines analyzed, while the expression of PC5B is limited. These PCs can cleave their substrates within the constitutive secretory pathway

PCs GROUP	GROUP I	GROUP II	GROUP III	GROUP IV
TABLE 4: Cell and tissue distribution of PCs.				
PCs	FurinPACE/ SPC1/PCSK3 PC5B/PCSK6B PC7/LPC/PC8/ SPC7/PCSK6	PC1/3/PCSK1 PC2/PCSK2	PC5A/PCSK6A PACE4/PCSK5	PC4/PCSK4
Tissue distribution	Ubiquitous	Neuroendocrine	Ubiquitous	Germinal cells
Cell localization	Network-Trans Golgi, endosome, cell surface	Secretion granules	Network-Trans Golgi, secretion granules, cell surface	Undetermined

and/or cell surface (Figure 4). The second group includes PC1/3 and PC2 (Table 4). These convertases are exclusively expressed in neuroendocrine cells and tissues and are involved in the cleavage of substrates located in the granules of the regulated secretory pathway (Figure 4). The third group consists of PACE4 and PC5A that are expressed in both endocrine and non-endocrine cells (Table 4). These PCs cleave their substrates in the regulated secretory pathway and the constitutive secretory pathway (Figure 4). The fourth group that includes only PC4 whose expression was observed exclusively in germ cells [18, 39].

1.2.1 Convertases of Group I

Furin is the best-studied and most-characterized PC. As for PC7, the tissue distribution of Furin was mainly analyzed by Northern blot and/or in situ hybridization. These analyses show that these two PCs are expressed in all mammalian cells analyzed so far [40]. Furin is found in large quantities in the liver and kidney [41], while PC7 is mainly expressed in lymphoid tissues such as the thymus, T cells, and spleen [24, 41]. The expression of PC5B is more limited with high expression in the intestine (jejunum, duodenum, ileum, colon), the kidney, and the liver [21, 42]. Active forms of Furin [43–44], PC7 [45], and PC5/6B were found to be mainly located to the TGN, endosomes, and on the cell surface. Indeed, it has been shown that the location and movement of Furin requires the presence of signals on its intracytoplasmic domain [43–44, 46]. Thus, phosphorylation of dif-

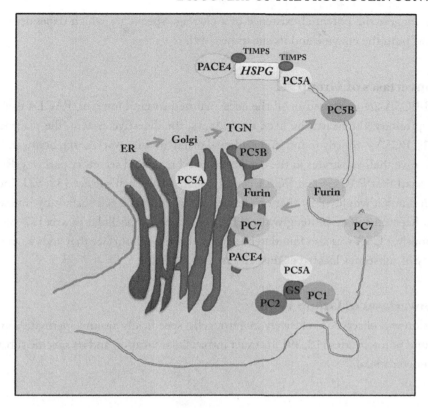

FIGURE 4: Schematic representation of the cellular localization of PCs. ER: Endoplasmic reticulum. TGN: Trans-Golgi-Network GS: granule secretion.

ferent patterns located in this area could lead to Furin progression to TGN [47] or to the plasma membrane via endosomes [48]. A truncated form of active Furin was found to be secreted from the plasma membrane and able to mediate the cleavage of molecules of different origins, synthesized and released as uncleaved precursors.

1.2.2 Convertases of Group II

This group contains PC1 and PC2 which are exclusively expressed in neural and endocrine cells, suggesting a possible specific role in these tissues. Indeed, these PCs were found to be expressed in the pituitary, hypothalamus, hippocampus, and cerebral cortex, and more recently, they have been localized in endocrine cells of the small intestine [49]. PC1 is found in greater quantities in the adenohypophysis, while PC2 is more abundant in the intermediate lobe of the pituitary [50]. The level of PC1 in endocrine cells of the small intestine is higher as compared with that of PC2. These observations suggest the potential existence of a cleavage specificity which depends on the tissue distribution of both the enzyme and its substrates [49].

1.2.3 Convertases of Group III

PACE4 and PC5A are expressed in all the cells analyzed so far. However, PACE4 is abundant in the anterior pituitary, the heart, the liver, the kidneys, the digestive system, the placenta, and the brain [40, 51]. PC5A is mostly found in the adrenal gland, uterus, ovary, aorta, brain, and lung [42]. These PCs cleave their substrates in the constitutive and regulated secretory pathway. Recent studies demonstrated that PACE4 and PC5A can be found on the cell surface [38, 52]. Indeed, their cysteine-rich domain would allow them to remain attached to the cell surface by interacting with TIMP-2 via heparan sulfate proteoglycans (HSPGs) of the extracellular matrix [37–38, 52] (Figure 4). Previously, PC5A was also found to be activated at the cell surface that adds a new regulation of the activity of substrates located at the cell surface [38, 53].

1.2.4 Convertases of Group IV

PC4 expression was observed exclusively in germ cells, specifically around spermatids and in macrophages found in the ovaries [18, 39]. Its exact intracellular location and its specific substrates have not yet been determined.

. . . .

CHAPTER 2

Substrates of the Proprotein Convertases

2.1 CLEAVAGE SITES OF THE PROPROTEIN CONVERTASES

Various studies have shown that PCs cleave their substrates at a monobasic, dibasic, or multi-basic residue. Previously, the group of Remacle et al. [54] conducted analysis regarding the substrates cleavage preference of each PC. This study was conducted by analyzing the ability of each PC to digest 100 synthetic decapeptides containing the PC cleavage sites RXR/K/XR. The analysis revealed that the presence of certain residues at specific positions seemed to influence the cleavage specificity of each PC [54]. The results of this study allowed the distribution of PCs according to the degree of their substrates cleavage preference. Thus, the first group contains only Furin; the second PC4, PC5/6, PC7, and PACE4; and third group only PC2.

2.2 CLASSIFICATION OF THE PROPROTEIN CONVERTASES SUBSTRATES

Based on the nature of the site recognized by PCs, various types of substrates have been identified and classified into four groups.

2.2.1 Substrates of Type I

The substrates of type I include all pro-proteins able to be processed in the constitutive secretory pathway and contain in their sequence the cleavage site RX-(K/R). The cleavage usually occurred at positions P4 and P2–P1. Precursors belonging to this group are mainly growth factors, receptors, bacterial toxins, and viral glycoproteins [55, 56] (Table 5).

2.2.2 Substrates of Type II

The precursors of this group are cleaved downstream from a pair of basic residues type (R/K)–(R/K). They consist mainly of precursor hormones such as insulin, PTH, as well as neuropeptides such as POMC (proopiomelanocortin) and several types of receptors (Table 6).

TABLE 5: Certain type I precursors that contain a cleavage site recognized by the PC

Precursors	Cleavage sequence							

Type I Precursors	P6	P5	P4	P3	P2	P1	P'1	P'2
	X -	X -	R -	X -	K/R -	R	X -	X

Growth factors and hormones

hPro-NGF	Thr – His – **Arg** – Ser – **Lys** - **Arg** ↓						Ser - Ser
hPro-BDNF	Ser –Met– **Arg** – Val – **Arg** - **Arg** ↓						His - Ser
hPro-NT3	Thr – Ser – **Arg** - Arg - **Lys** –**Arg** ↓						Tyr - Ala
hPro-NT4/5	Ala – Asn – **Arg** – Ser – **Arg** – **Arg** ↓						Gly - Val
hPro-PDGF-A	Pro – Ile – **Arg** – Arg – **Lys** – **Arg** ↓						Ser - Ile
hPro-PDGF-B	Leu – Ala – **Arg** – Gly – **Arg** – **Arg** ↓						Ser - Leu
hPro-TGFβ1	Ser – Ser – **Arg** – His – **Arg** – **Arg** ↓						Ala – Leu
Pro-Hepcidine	Gln – Arg – **Arg** – Arg – **Arg** – **Arg** ↓						Asp – Thr
Pro-Endothéline	Leu - Arg – **Arg** – Ser - **Lys** – **Arg** ↓						Cys - Ser

Growth factor receptors

Récepteur de l'Insuline	Pro - Ser – **Arg** – Lys – **Arg** – **Arg** ↓						Ser - Leu
Récepteur de l'IGF 1	Pro – Glu – **Arg** – Lys – **Arg** – **Arg** ↓						Asp - Val
Intégrine α3	Pro – Gln – **Arg** – Arg – **Arg** – **Arg** ↓						Gln - Leu
Intégrine α6	Asn – Ser – **Arg** – Lys - **Lys** – **Arg** ↓						Glu - Ile
Intégrine α7	Arg – Asp – **Arg** – Arg – **Arg** – **Arg** ↓						Glu - Leu
Intégrine αIIb	His – Lys - **Arg** – Asp – **Arg** – **Arg** ↓						Gln - Ile
Récepteur de la Leptine	Gln – Val – **Arg** – Glu – **Lys** – **Arg** ↓						Leu - Asp
Récepteur de Notch I	Gly – Gly – **Arg** – Gln – **Arg** – **Arg** ↓						Glu - Leu

Serum proteins

Pro-von Willebrand factor	Ser – His – **Arg** – Ser – **Lys** – **Arg** ↓						Ser - Leu
Pro-Factor IX	Leu – Asn - **Arg** – Pro – **Lys** – **Arg** ↓						Tyr - Asn
Pro-Factor X	Lys – Glu – **Arg** – Arg – **Lys** – **Arg** ↓						Ser - Val

TABLE 5: (*continued*)

Proteases

ADAM-9	Leu – Leu – **Arg** - Arg –**Arg** – **Arg** ↓	Ala - Val
ADAM-10	Leu – Leu – **Arg** – Lys – **Lys** – **Arg** ↓	Thr – Thr
ADAM-12	Ala – Arg – **Arg** – His – **Lys** – **Arg** ↓	Glu - Thr
ADAM-15	His – Ile – **Arg** – Arg – **Arg** – **Arg** ↓	Asp - Val
ADAM-17	Val – His - **Arg** – Val – **Lys** – **Arg** ↓	Arg – Ala
ADAMTS-1	Ser – Ile – **Arg** – Lys – **Lys** - **Arg** ↓	Phe - Val
ADAMTS-13	Arg – Gln – **Arg** – Gln – **Arg** – **Arg** ↓	Ala - Ala
MT1-MMP	Asn –Val – **Arg** – Arg – **Lys** – **Arg** ↓	Tyr - Ala
MT2-MMP	His – Ile – **Arg** – Arg – **Lys** – **Arg** ↓	Tyr – Ala
MT3-MMP	Gln – Ala – **Arg** – Arg – **Arg** – **Arg** ↓	Gln - Ala

Glycoproteins

Gp 160 de HIV-I	Val – Gln – **Arg** – Glu – **Lys** – **Arg** ↓	Ala - Val
GP du virus Ebola	Gly – Arg – **Arg** – Thr – **Arg** – **Arg** ↓	Glu - Ala
GP du virus EBV	Lys – Arg – **Arg** – Arg – **Arg** – **Arg** ↓	Glu - Ala
Toxine de la diphtérie	Gly – Asn – **Arg** – Val – **Arg** –**Arg** ↓	Ser - Val
Anthrax PA83	Asn – Ser – **Arg** - Lys - **Lys** - **Arg** ↓	Ser - Thr

2.2.3   Substrates of Type III

Type III substrates include substrates cleavable at monobasic sites occupied by a single basic residue K/R. Generally, the cleavage of these substrates requires the presence of basic residues at positions P4, P6, P8 or located upstream from the cleavage site [17, 57]. In this group, we found several growth factors, such as IGF-1, IGF-2, and EGF (Table 7).

2.2.4   Substrates of Type IV

These substrates include precursors that can be cleaved at monobasic or dibasic residues. This group differs from the other groups because it requires the presence of a basic residue (R/K) downstream from the cleavage site (P2'), in addition to the presence of basic residues in positions P4, P6, or P8 (Table 8).

TABLE 6: Certain type II precursors that contain a cleavage site recognized by the PC

Type II Precursors	P6	P5	P4	P3	P2	P1	P'1	P'2
	X -	X -	X -	X -	K/R -	R	X -	X

Growth factors and hormones

Pro-PTH	Lys –	Ser –	Val –	Lys –	**Lys –**	**Arg** ↓	Ser -	Val
ProInsuline (chaîne) B/C	Thr –	Pro –	Lys –	Thr –	**Arg –**	**Arg** ↓	Glu -	Ala
(chaîne) C/A	Gly –	Ser –	Leu –	Gln –	**Lys –**	**Arg** ↓	Gly -	Ile
Pro-Gastrine	Ala –	Ser –	His –	His –	**Arg –**	**Arg** ↓	Gln -	Leu
VEGF-C	His –	Ser –	Ile –	Ile –	**Arg –**	**Arg** ↓	Ser –	Leu
VEGF-D	Tyr –	Ser –	Ile –	Ile –	**Arg –**	**Arg** ↓	Ser -	Ile
POMC (MSH/CLIP)	Pro –	Val –	Gly –	Lys –	**Lys –**	**Arg** ↓	Arg –	Pro
(LPH/END)	Pro –	Pro –	Lys –	Asp –	**Lys –**	**Arg** ↓	Tyr –	Gly
(ACTH/βLPH)	Pro –	Leu –	Glu –	Phe –	**Lys –**	**Arg** ↓	Glu –	Leu
Pro-Encéphaline (142/143)	Gly –	Gly –	Phe –	Met –	**Lys –**	**Arg** ↓	Asp -	Ala
(195/196)	Met –	Asp –	Tyr –	Gln –	**Lys –**	**Arg** ↓	Tyr –	Gly
(236/237)	Gly –	Gly –	Phe –	Leu –	**Lys -**	**Arg** ↓	Phe -	Ala
Pro-Rénine	Ser –	Gln –	Pro –	Met –	**Lys –**	**Arg** ↓	Leu -	Thr
Pro-Neurotensine	Pro –	Tyr –	Ile –	Leu –	**Lys –**	**Arg** ↓	Ala -	Ser

Receptors

Intégrine a4	His –	Val –	Ile –	Ser –	**Lys -**	**Arg** ↓	Ser -	Thr
Intégrine a5	His –	His –	Gln -	Gln –	**Lys -**	**Arg** ↓	Glu –	Ala
Intégrine av	His –	Leu –	Ile –	Thr –	**Lys -**	**Arg** ↓	Asp -	Leu
Intégrine a8	His –	Tyr –	Ile –	Arg –	**Arg -**	**Arg** ↓	Glu -	Val

Serum proteins

Pro-Albumine	Arg –	Gly –	Val –	Phe –	**Arg –**	**Arg** ↓	Asp -	Ala
Pro-Protéine C	Arg –	Ser –	His –	Leu –	**Lys –**	**Arg** ↓	Asp -	Thr

TABLE 7: Certain type III precursors that contain a cleavage site recognized by the PC

Type III Precursors	P6	P5	P4	P3	P2	P1	P'1	P'2
	X -	X -	X -	X -	X-	K/R	X –	K/R

Growth factors and hormones

	P6	P5	P4	P3	P2	P1	P'1	P'2
Pro-IGF-I	Pro –	Thr –	**Lys** –	Ala –	Ala –	**Arg** ↓	Ser -	Ile
Pro-IGF-II	Pro –	Ala –	**Lys** –	Ser –	Glu –	**Arg** ↓	Asp -	Val
Pro-EGF(C-ter)	**Arg** –	Trp –	Trp –	Glu –	Leu –	**Arg** ↓	His –	Ala
Pro-FGF-23	Pro –	Arg –	**Arg** –	His –	Thr –	**Arg** ↓	Ser –	Ala
Pro-PDGF-C	Phe –	Gly –	**Arg** –	Lys –	Ser –	**Arg** ↓	Val –	Val
Pro-PDGF-D	His –	Asp –	**Arg** –	Lys –	Ser –	**Arg** ↓	Val -	Asp

Proteases

	P6	P5	P4	P3	P2	P1	P'1	P'2
Pro-MMP-1	Val –	Met –	**Lys** –	Gln –	Pro -	**Arg** ↓	Cys -	Gly
Pro-MMP-8	Met –	Met –	**Lys** –	Lys –	Pro –	**Arg** ↓	Cys -	Gly
Pro-MMP-13	Val –	Met –	**Lys** –	Lys –	Pro –	**Arg** ↓	Cys -	Gly

TABLE 8: Certain type IV precursors that contain a cleavage site recognized by the PC

Type IV Precursors	P6	P5	P4	P3	P2	P1	P′1	P′2
	X -	X -	X -	X -	X -	K/R	X –	K/R

Growth factors and hormones

Pro-Glucagon	Leu –	Met –	Asn –	Thr –	**Lys** –	**Arg**↓	Asn –	**Arg**
Pro-MIS	**Arg** –	Gly –	**Arg** –	Ala –	Gly –	**Arg**↓	Ser -	**Lys**

Receptors

Pro-récepteur PTPu	Glu –	Glu –	**Arg** –	Pro –	**Arg** –	**Arg** ↓	Thr -	**Lys**

CHAPTER 3

Inhibitors of the
Proprotein Convertases

The involvement of various PC substrates in the regulation of several physiological and pathological processes has led to the development of various inhibitors of PCs. These PC activity regulators can be divided into two groups, namely, endogenous and exogenous inhibitors. Endogenous inhibitors of PCs are ProSaas, 7B2, and PCs prosegments. The group of exogenous inhibitors includes chemical inhibitors, peptides, and protein based-inhibitors. An overview of various exogenous PCs inhibitors is discussed in the contribution authored by Ajoy Basak in this same Colloquium e-book Series: http://www.morganclaypool.com/doi/abs/10.4199/C00066ED1V01Y201209PAC003. In this chapter, only the features and the importance of α1-PDX as a general exogenous inhibitor of the PCs is described.

3.1   ENDOGENOUS INHIBITORS OF THE PCs
3.1.1   The Naturally Occurring Inhibitor of PC2
Braks and Martens were the first to show that 7B2 acts as a chaperone for proPC2. The interaction "7B2-proPC2" seemed to occur only in the early compartment of the secretory pathway and dissociate from it in the latter ones [42, 85–60]. Lately, it was proposed that 7B2 facilitates proPC2 transport from the ER to the secretory granules and participates in the generation of fully active PC2 [60]. In the presence of an alkaline pH, pro7B2 and proPC2 interact in the ER and form an inactive complex (pro7B2-proPC2). During its progression through the TGN in the presence of decreased pH and increased $[Ca^{2+}]$, the pro7B2 is cleaved by the PCs and release a C-terminal fragment with inhibitory function on PC2. Additional pH decreases and $[Ca^{2+}]$ increases in the secretory granule-induced proPC2 auto-activation and liberates the prodomain of PC2 which generates active PC2. Subsequently, the N-terminal domain of PC2 and C-terminal domain of 7B2 are rapidly degraded by PC2 and carboxypeptidase E (Figure 5).

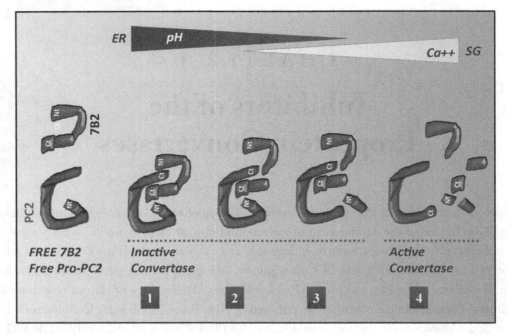

FIGURE 5: PC2 activity regulation by 7B2. Under alkaline pH, ProPC2 interact with Pro7B2 to form an inactive complex (pro7B2-proPC2) (1). During the complex progression in the trans-Golgi network where gradually the pH is decreased and $[Ca^{2+}]$ is increased, pro7B2 is cleaved by the PC, while the C-terminal fragment of 7B2 remained attached to proPC2 (2). In the secretory granule (SG), the proPC2 is cleaved auto-catalytically to liberate the prodomain fragment (3). The N-terminal domain of PC2 and the C-terminal and N terminal domains of 7B2 are rapidly degraded by PC2 and carboxypeptidase E and generated a fully active PC2 (4).

3.1.2   ProSAAS: The Naturally Occurring Inhibitor of PC1

The sequence L–L–R–V–K–R, located in the proSAAS C-terminal domain, seemed to be responsible for the inhibitory potency of PC1 [61]. This inhibitor was identified by combinatorial library peptide screening as a tight binding site for PC1 [62]. In the ER and Golgi apparatus, PC1 is inactive due to the neutral pH and relatively low Ca^{2+} levels in addition to its interaction with proSAAS. Following its progression through the TGN, proSAAS is cleaved into two peptides designed as PEN and LEN fragments that remove the inhibition of PC1 by proSAAS (Figure 6).

3.1.3   Prosegment of Proprotein Convertases

To date, the prosegments of PCs are the only known naturally occurring inhibitors of the PCs found in the secretory pathway. Several studies have shown that the prosegment acts as an intracellular

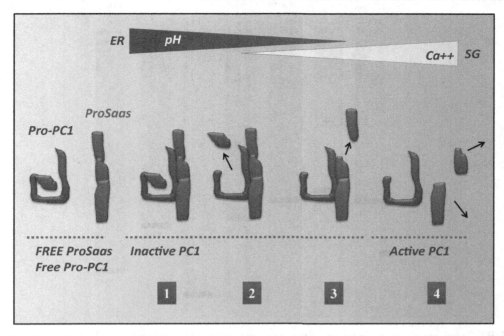

FIGURE 6: PC1 activity regulation by ProSaas. In the ER, proPC1 form with ProSaas an inactive complex (proPC1-ProSaas) (1). The progression of this complex to the trans Golgi network resulted in the first cleavages of ProSaas (2) and ProPC1 (3). At this step the C-terminal fragment of Prosaas is attached and inhibits PC1. In the secretory granules the ProSaas is cleaved into two peptides designed as PEN and LEN fragments and liberate an active PC1 (4).

chaperone protein for correct folding of proteins, their transport, and secretion. The prosegment also acts as a modulator of PCs activity. This regulation is controlled by the variation of pH and calcium concentration [63]. Indeed, as their substrates, PCs are also synthesized as inactive precursors and activated during their progression from the ER to the TGN by autocatalytic cleavage (Figure 3) which releases a segment of 80–100 amino acids (Figure 7). Although each segment is an inhibitor of its own enzyme, all PC prosegments can also act as potent inhibitors for various PCs [63–66]. For example, the prosegment of Furin is 10 times more potent on the PC5 enzyme (IC_{50} 0.4 nM) as compared to its action on its own enzyme Furin, while the prosegment of PC7 is relatively selective for its enzyme (IC_{50} 0.4 nM) [67] (Table 9). Previously, these prosegments were found to be able to inhibit the maturation and function of various substrates of the convertases involved in the malignant phenotype of tumor cells, such as VEGF-C [68], PDGF-A [69], PDFG-B [70], IGF 1R MT1-MMP-1 [71–72], and TGF [73].

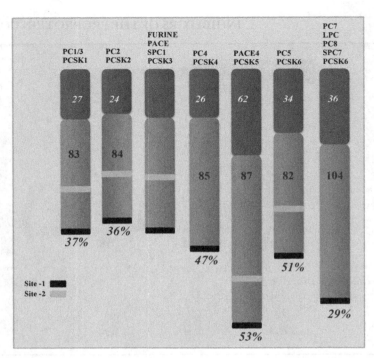

FIGURE 7: Schematic representation of the structure of the PC prosegments. The figures indicate the percentage in homology in the prosegment of each convertase as compared to that of the Furin. The number of the amino acids and the two cleavage sites of PC prosegments are also shown.

TABLE 9: IC50 and Ki values of PC prosegments. Some PC prosegments are able to inhibit various PCs			
PROSEGMENTS OF PCs	**PCs**	**IC50**	**KI**
ppPC1	**Furin**	**ND**	**>1000 nM**
	PC5	**ND**	**>1000 nM**
	PC7	**ND**	**>1000 nM**
ppPC2	**Furin**	**ND**	**>1000 nM**
	PC5	**ND**	**>1000 nM**
	PC7	**ND**	**>1000 nM**
ppFurin	**Furin**	**4 nM**	**ND**
	PACE4	**110 nM**	**ND**
	PC5	**0.4 nM**	**ND**
	PC7	**20 nM**	**ND**

PROSEGMENTS OF PCs	PCs	IC50	KI
ppPC4	Furin	ND	180 nM
	PC5	ND	23 nM
	PC7	ND	117 nM
ppPC5	Furin	ND	152 nM
	PC5A	ND	27 nM
	PC7	ND	926 nM
ppPC7	Furin	40 nM	>1000 nM
	PACE4	25 nM	ND
	PC5	20 nM	ND
	PC7	0.4 nM	ND

TABLE 9: (*continued*)

3.2   EXOGENOUS INHIBITOR OF THE PCs: THE α1-ANTITRYPSIN VARIANT (α1-PDX)

The discovery of this inhibitor was initially based on observations made on a patient who had an inability to convert the pro-albumin into albumin due to a mutation of α1-antitrypsin that inactivate the enzyme [74–75]. The α1-antitrypsin is a serpin (serine protease inhibitor) responsible for the physiological inhibition of the neutrophil elastase. The serpins present the particularity to form a stable complex with the enzyme, essential for the process of enzyme inactivation [76]. In fact, they act as a suicide inhibitor by interacting with protease at their reactive loop (RSL) in the active site. Thus, after cleavage by the protease, the serpin undergoes a conformational change and forms a stable complex with the enzyme [77]. A mutation in the active site of the serpin alters its specificity. Thereby, mutation of the natural mutation located at the active site of α1-antitrypsin (AIPR^{358} to AIPM^{358}) generates a potent inhibitor of thrombin called α1-AT Pittsburg (α1-PTT) (Figure 8) [74]. Subsequently, based on the structure of the cleavage site of Furin found in various substrates, the group of Anderson et al. introduced a second mutation in α1-PTT, changing the A^{355} residue to R^{355} (AIPR in IRP) in position P4. This mutation not only helped to create a site recognizable by the convertases (RXXR) but also to create a potent inhibitor of Furin. This variant was called α1-antitrypsine Portland or α1-PDX (Figure 8) [78]. During its interaction with the enzyme, α1-PDX cleaved by the PC undergoes a conformational change and form a stable complex with the protease that is degraded intracellularly (Figure 9).

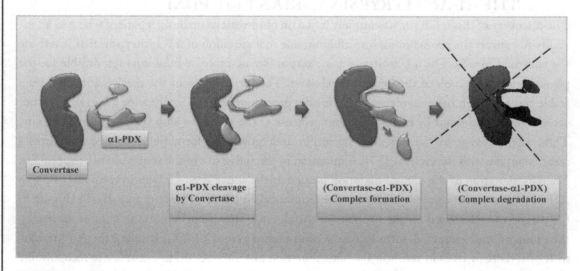

First mutation transform α1-AT to α1-AT PIT

α1-AT ^{341}EKGTEAAGAMFLEAIPM^{358} SIPPEVKFNKPJVFLMI^{375}

α1-AT PIT ^{341}EKGTEAAGAMFLEAIP<u>R</u>^{358} ↓ SIPPEVKFNKPJVFLMI^{375}

Second mutation transform α1-AT PIT to α1-PDX

α1-AT PIT ^{341}EKGTEAAGAMFLEAIP<u>R</u>^{358} ↓ SIPPEVKFNKPJVFLMI^{375}

α1-PDX ^{341}EKGTEAAGAMFLE<u>R</u>IP<u>R</u>^{358} ↓ SIPPEVKFNKPJVFLMI^{375}

FIGURE 8: The variant of α1-antitrypsin (α1-PDX) as an inhibitor of PCs. The first mutation (AIPM^{358} into AIPR^{358}) of α1-antitrypsin (α1-AT1) generates the Pittsburg α1-antitrypsin (α1AT PIT). The introduction of a second Arg (A^{355}IPR in R^{355}IPR) creates a potent inhibitor of Furin called α1-PDX. The arrows indicate the cleavage sites.

FIGURE 9: Mechanism of inhibition of PCs by α1-PDX. α1-PDX acts as a PCs substrate during its interaction with the convertase. The complex formed is thereby dissociated and degraded intracellularly.

The specificity of PCs inhibition by α-1-PDX is controversial. Indeed, previous *in vitro* studies demonstrated that this inhibitor seemed to mainly act in the constitutive secretory pathway and inhibit all the convertases found in this pathway [79–80]. Subsequently, the group of G. Thomas revealed that α1-PDX inhibits specifically Furin and to a lesser extent PC5B [81], PC7, and PACE4 [1]. Lately, α1-PDX was also found able to inhibit PACE4 forming a stable complex similar to that formed during its interaction with Furin [82]. However, Tsuji et al. demonstrated using a rat α-1-antitrypsin that it was possible to make an α-1-PDX more selective toward Furin, PC5, and PACE4 following mutation of the residues Pro and Met to two Arg residues. Subsequently, other mutations in α1-PDX sequence resulted in a more selective inhibitor capable of inhibiting PC5 and PACE4 but not Furin [82].

Early studies of α-1-PDX expression in various cells demonstrated that this inhibitor was able to prevent the maturation of various viral glycoproteins such as gp160 of HIV [78] and glycoprotein F of measles virus [83]. Subsequently, a large number of proproteins maturation were found to be inhibited by α1-PDX. These include various growth factors (TGF-b, VEGF-C, PDGF-A, etc.), integrins, MMPs, and many others PCs substrates involved in the malignant phenotype of tumor cells [2].

·   ·   ·   ·

CHAPTER 4

Proprotein Convertases in Physiology and Pathology

4.1 PHYSIOLOGICAL ROLE OF THE PROPROTEIN CONVERTASES

PCs are involved in many physiological processes, as evidenced by the phenotypes observed in various PC knockout mice that emphasize the complexity of the role of these enzymes and the wide range of their substrates and downstream effectors [84]. Despite the fact that several precursors can be cleaved by different PCs, the phenotypes generated by the silencing of each PC demonstrated the potential existence of specific substrates for each PC whose maturation is essential at least during embryonic development.

To date, the information gathered on the various PC knockout mice show that the absence of Furin gene (*fur, PCSK3*) or PC5/6 (*PCSK5*) is lethal early in embryonic development, suggesting a key role of these PCs during embryogenesis. The mouse knockout of Furin induces neural and cardiac abnormalities, and PC5/6 (PCSK5) silencing induces embryonic lethality during the processes of implantation [42] (Figure 10). The knockout mice of PC1 (*PCSK1*) or PC2 (*PCSK2*) genes are viable and show various metabolic and endocrine disorders. In addition, knockout mice of PC1 (*PCSK1*) present developmental abnormalities while that of PC2 (*PCSK2*) causes growth retardation. The knockout mouse for PACE4 (*PCSK6*) is viable at 75% with craniofacial abnormalities. The absence of the gene PC4 (*PCSK4*) results in decreased fertility while the mouse knockout of PC7 (*PCSK7*) seems to not show any apparent abnormalities (Figure 10) [84].

4.2 LACK OF PROPROTEIN CONVERTASES ACTIVITY IN HUMANS

In humans, only the absence of PC1 activity has been described so far. Indeed, previous studies have shown a direct link between the mutation of the PC1 gene and physiological dysfunction in humans. Initially, the group O'Rahilly described the case of a patient with various hormonal disorders, such as reactive hypoglycemia, adrenal insufficiency of partial primary amenorrhea, and

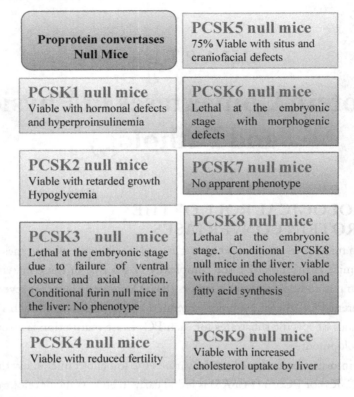

Proprotein convertases Null Mice

PCSK5 null mice
75% Viable with situs and craniofacial defects

PCSK1 null mice
Viable with hormonal defects and hyperproinsulinemia

PCSK6 null mice
Lethal at the embryonic stage with morphogenic defects

PCSK2 null mice
Viable with retarded growth Hypoglycemia

PCSK7 null mice
No apparent phenotype

PCSK3 null mice
Lethal at the embryonic stage due to failure of ventral closure and axial rotation. Conditional furin null mice in the liver: No phenotype

PCSK8 null mice
Lethal at the embryonic stage. Conditional PCSK8 null mice in the liver: viable with reduced cholesterol and fatty acid synthesis

PCSK4 null mice
Viable with reduced fertility

PCSK9 null mice
Viable with increased cholesterol uptake by liver

FIGURE 10: Summary of phenotypes found in each PC knockout mice.

childhood obesity [85]. Hormonal analysis revealed that plasma showed a large amount of non-matured prohormones, such as proinsulin and POMC [85], suggesting the existence of a lack of maturation processes in this patient. Indeed, this hypothesis was confirmed by the group of Jackson et al. who revealed that this patient in fact present two mutations, one in each allele of *PCSK1* gene. The first mutation was responsible for an abnormal transcript resulting in a truncated protein at the catalytic site. The other mutation was a point mutation of proPC1, making it unsuitable for maturation and subsequently causing its retention in the ER [86]. Later, the case of a second patient was described. This patient was a newborn with severe diarrhea and several postnatal abnormalities. The analysis of the coding sequence of PC1 revealed the presence of two mutations, one in each allele. The mutations were not identical to those found in the first patient. However, the first mutation gave rise to a truncated protein at the catalytic site, while the second induces a change in a highly conserved residue at the catalytic site which rendered the enzyme inactive [87]. Later, a detailed study in the first patient revealed that she had alternating constipation and diarrhea. Other studies have shown that the newborn suffered from defective absorption of monosaccharides and lipids, suggesting the involvement of PC1 in intestinal absorption by activating substrates regulators involved in this process [87]. Lately, a third patient with a mutation in the gene *PCSK1* was reported by the same group. The patient was also found to present obesity and persistent diarrhea but no hypoglycemia was detected [88].

Lately, other studies revealed a correlation between polymorphisms in the gene *PCSK1* and obesity. Indeed, the analysis of this gene in 13,659 individuals identified the presence of several variants of *PCSK1* in obese adults and children. Further analysis revealed the enzymatic activity of the enzyme was affected in only one PCSK1 variant [89].

4.3   ROLE OF PROPROTEIN CONVERTASES IN PATHOLOGY

4.3.1   Neurodegenerative Diseases

PCs have been linked to some neurodegenerative disorders via their direct or indirect roles in the production of amyloidogenic peptides. In Alzheimer's disease, the amyloid-β (Aβ) is the principal component of senile plaques. The latter is generated by proteolytic cleavage of its precursor by β- and γ-secretases. Previously, the PCs were found to process the zymogens of these two enzymes, thereby directly implicating the PCs in this disease [4, 90].

4.3.2   Bacterial Toxins and Proprotein Convertases

All three classes of bacterial toxins are activated by PCs.

Toxins of the first class are characterized by their polypeptide chain containing both the toxic unit and the unit responsible for the interaction with the target cell. In this class, the toxins are synthesized as precursor proteins and cleaved by PCs during their interaction with the cell surface or endosomal compartments of the target cell. This class includes the *Diphtheria* toxin [91], *Pseudomonas aeruginosa* exotoxin A (PEA) [92], *Botulinum* neurotoxin, and *Bordetella* dermonecrotic toxin [93].

The second class of toxins such as anthrax *Bacillus anthracis* are synthesized as separate polypeptide chains that interact together at the cell surface after activation by the PCs [94–95].

The last class, which includes the pore-forming toxins, such as of the *Aeromonas hydrophila* aerolysin toxins, are produced and secreted in the form of dimers that bind to glycosylphosphatidyl inositol of the cell membrane. Generally, the activation of these toxins by the PCs occurs at the surface of the target cells [96].

4.3.3   Viral Infection and Proprotein Convertases

Studies on various viruses revealed that the cleavage of different glycoproteins of the viral envelope by one or more PCs is indispensable for acquiring infectious viral particles. These include the viral glycoproteins of HIV-1 [79], Hong Kong virus, Ebola, and SARS [97, 98]. Some of these studies have shown that inhibition of the maturation of the viral glycoproteins by inhibitors of PCs, such as dec-CMK-RVKR, block the ability of these viruses to infect the host cell.

CHAPTER 5

Proprotein Convertases and Tumorigenesis

Initially, the involvement of PCs in carcinogenesis and tumor progression was suggested based on the role of these enzymes in the activation and expression of several proteins directly involved in neoplasia [2]. Among these molecules, there are various growth factors, their receptors, adhesion molecules, and metalloproteases [2]. The inhibition of the activity of PCs in various tumor cell lines was reported to significantly affect the malignant phenotype of tumor cells. In addition, the use of site-directed mutagenesis revealed the importance of the maturation of certain substrates of the PCs, such as VEGF-C, IGF-1R, MT1-MMP [5, 99], in the mediation of their cellular functions. These include tumor cells proliferation, survival, migration, and invasion [2, 5]. In 2001, our team reported for the first time direct evidence regarding the involvement of PCs in carcinogenesis [72].

5.1 PROTEIN MATURATION IN TUMOR PROGRESSION AND METASTASIS

There are over a hundred forms of cancer, all characterized by uncontrolled growth of tumor cells. Tumor progression is a complex biological process in which various cellular functions, such as cell proliferation, differentiation, adhesion, and migration, were disordered. Initially, tumor growth is limited by the lack of space and lack of supply of oxygen and nutrients. Thus, to overcome these constraints, other cells were found to be recruited into the tumor tissue, such as endothelial cells (EC) and inflammatory cells (IC). Indeed, the ECs ensure the development of new blood vessels and improve the supply of oxygen and nutrients and removal of waste, while IC provide the invading tumor factors that facilitate tumor growth. All of these processes allow tumor cells to continue to grow locally and/or diffuse into the surrounding tissue and, in some cases, to establish a secondary site and form metastases.

Metastasis formation begins with the detachment of one or a group of tumor cells from the primary tumor, and then they migrate and invade the surrounding tissue. After the invasion, the tumor cells enter and circulate in the blood or lymph vessels prior their attachment to the endothelium in order to proliferate, and form metastases. The formed lesion can in turn become a source of

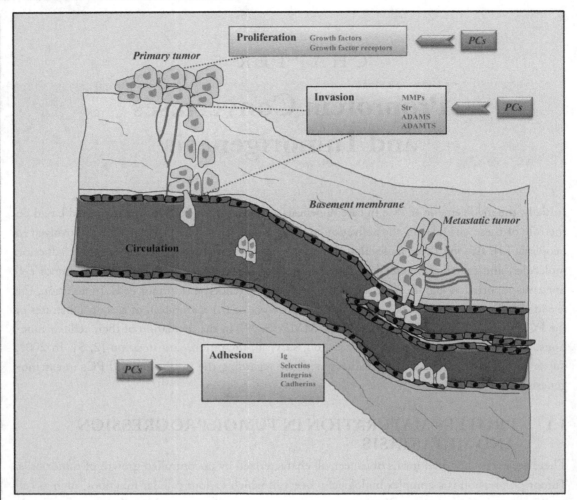

FIGURE 11: Schematic representation of PCs role in the regulation of processes involved in tumor progression and the formation of metastases. Activation or induction of the expression by PCs of several growth factors and/or their receptors control tumor cell proliferation. Activation/induction of adhesion molecules by PCs influence tumor cells adhesion, invasion, and migration and subsequently metastases formation. Arrows indicate the potential sites for PCs inhibitors.

dissemination of tumor cells and give rise to a secondary metastasis, a main morbid complication associated with cancer. Recently, it has been shown that several molecules involved in these processes are proteolytically activated by the PCs. Thereby, by activating these molecules, the PCs can directly or indirectly control the malignant phenotype and metastatic potential of tumor cells, such as their proliferation, survival, invasion, and migration [2, 100] (Figure 11). In addition, several studies have shown that different PCs were strongly expressed in various types of cancer [101], such as breast [102], lung [103], and neck and head cancer [104].

5.2 SUBSTRATES OF PROPROTEIN CONVERTASES AND THE MALIGNANT PHENOTYPE GROWTH FACTORS

5.2.1 Growth Factors

Under physiological conditions, cell proliferation is generally provided by growth factors. In contrast, the loss or reduction of the necessity of these factors is a process common to all tumor cells. However, over-expression and/or increased activity of these molecules plays a crucial role in maintaining the transformed character and/or metastatic potential of tumor cells by increasing their uncontrolled proliferation and invasion [2].

All growth factors can be grouped into two categories: growth factors called "competent factors," such as PDGF, VEGF, and bFGF. These molecules provide input cells in the G1 phase of the cell cycle. The other category assembles growth factors called "progression factors" such as IGF-1, which allows the passage of cells from G1 to S phase of the cell cycle and subsequently division [105–107]. A large number of these growth factors are synthesized as precursor proteins and then are matured and/or activated by PCs [2] (Table 5). Thus, in some cases, increased expression of these molecules and their activating proteases can significantly influence the proliferation of tumor cells. Among the growth factors described as substrates of PCs are IGF-1 [108], endothelin [109], NGF [110, 111], PTH [112], and TGF-β1 [113]. In addition, based on the presence of the cleavage site RXXR in their amino acid sequence, other growth factors were also suggested as potential substrates of PCs (Table 5).

The importance of the maturation of these molecules by PCs in the mediation of tumor cells proliferation was initially suggested due to the observed anti-proliferative effect of PC inhibition on several tumor cell lines [2]. Indeed, at that time, the inhibition of Furin was found to reduce the proliferative capacity of several types of tumor cells, such as Leydig cells H-500 [114], gastric GSM06 [115], and pancreatic min6 [116]. In addition, Kayo et al. found that the conditioned medium derived from min6 cells over-expressing Furin induced the proliferation of parental lines. Thereby, these observations already suggested that the over-expression of Furin in tumor cells generated molecules with a stimulatory effect on cell proliferation. Lately, in agreement with these results, the use of α1-PDX and site-directed mutagenesis, our team demonstrated the importance of the maturation of some substrates by the convertases in various cellular functions including cell proliferation, invasion, and survival. Among the molecules that we identified as substrates of the convertases, and their maturation was found necessary for the mediation of their function, are

PDGF-A [69], PDGF-B [70], VEGF-C [68], and IGF-1 R [72, 2]. Underneath is an example for the importance of growth factors (PDGF-A and PDGF-B) processing by the PCs is described. In this same Colloquium series, the chapter by G. Siegfried and AM Khatib will discuss the importance of VEGF-C and VEGF-D importance in tumorigenesis and angiogenesis.

ProPDGF-A and ProPDGF-B Precursors Processing. Using the PC activity-deficient colon carcinoma cell line LoVo, we found that Furin is the most potent PDGF-A convertase and mutation of PDGF-A PC cleavage site RRKR[86] to ARKA[86] inhibited pro-PDGF-A processing. *In vitro* inhibition of proPDGF-A maturation results in a loss ofits ability to phosphorylate PDGF receptor and induce cell proliferation. *In vivo*, in nude mice, subcutaneous injection of the Chinese Hamster Ovary (CHO) tumor cells in the cells over-expressing mutant pro-PDGF-A ARKA[86] failed to induce tumor formation [69]. Unlike the PDGF-A, a large portion of the PDGF-B is produced normally and is retained at the cell surface. The inhibition of pro-PDGF-B maturation by mutagenesis demonstrated the involvement of the PCs in this mechanism. Indeed, after its synthesis as precursor, pro-PDGF-B undergoes two proteolytic cleavages necessary for its secretion. The first cleavage is carried out by convertases, while the latter is ensured by an unknown enzyme. The cleavage site by the PCs is located upstream from the second site that contains the amino acid sequence involved in the retention of the PDGF-B on the cell surface (Figure 12). We have demonstrated that the cleavage of pro-PDGF-B by convertases at the first cleavage site is essential for proPDGF-B cleavage at the second site to take place. Thus, inhibition of convertase blocks the cleavage of both sites of PDGF-B and prevents its secretion [70].

5.2.2 Growth Factor Receptors

Most of the growth factors listed above transmit their signals through receptor tyrosine kinases. These transmembrane receptors, once activated by their ligands dimerize before auto-phosphorylation. This process results in the induction of a cascade of intracellular signals responsible for the induction of several biological responses including proliferation, survival, migration, and cell invasion. Several proteins involved in this cascade of phosphorylation were identified. These include phospholipase Grb [117], PLCγ [118], GTPase-activating protein [119], PI3K [120], IRS [121], SHC, and PKB/Akt [122]. Today, it is well established that several growth factor receptors using these kinases are involved in the malignant phenotype of several tumor cell lines and in tumor progression and metastasis. Indeed, over-expression and/or mutations/insertions in this type of receptor lead to constitutive activation of protein kinases [123]. Generally, this hyper-activation is frequently accompanied by constitutive expression of their respective ligands, thus providing tumor cells with maintained stimulatory action [124]. Like their ligands, some of these growth factor receptors con-

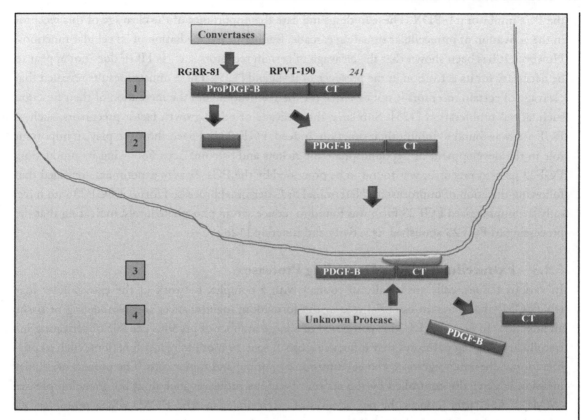

FIGURE 12: Maturation and secretion of PDGF-B. After its synthesis pro-PDGF-B is retained at the cell surface with basic residues located on the C-terminal (CT) immediately after the site ARPVT 190. The PDGF-B cleavage at this site depends on the PCs cleavage site RGRR81 to facilitate mature PDGF-B secretion.

tain in their amino acid sequence a cleavage site recognized by one or more PCs such as HGF-R [125], insulin receptor [126], and IGF-1R [71–72].

Previously, Hwang et al. showed that cleavage of the insulin receptor is essential for its activation and intracellular signal transmission [127]. We also showed that the maturation of IGF-1 R is essential for the activation and maintenance of the transformed character of tumor cells in the presence of IGF-1. [72]. Indeed, the inhibition of the maturation of this receptor by expression of α1-PDX in colon cancer cells blocks the ability of IGF-1 to induce the phosphorylation of IGF-1 R and its direct effector, IRS-1 [72]. Furthermore, IGF-1 failed to induce a significant effect on cell proliferation and protection against apoptosis in cells with unprocessed IGF-1R stably expressing the PCs inhibitor α1-PDX. These findings indicate the importance of the cleavage of this receptor in the activation of intracellular signaling cascade, leading to the mediation of its cellular functions. However, it has been shown that the cleavage of certain receptors such as HGF does not appear to be necessary for its activation in the presence of its ligand [125]. These unusual results revealed that cleavage of certain receptors is not essential for their activation and the mediation of their function such as cell proliferation [125]. Similarly, the cleavage of other growth factor precursors, such as FGF-23, was found to inhibit their function. Indeed, FGF-23 has been shown to play an important role in the development of hypophosphatemic rickets and osteomalacia. Following its production, FGF23 protein precursor was found to be processed by the PCs. *In vivo* experiments revealed that following injection of unprocessed, N-terminal or C-terminal processed form of FGF23 into mice, only the unprocessed FGF23 form was found to reduce serum phosphate levels, indicating that the processing of FGF23 abolished its activity and function [128].

5.2.3 Extracellular Matrix Degrading Proteases

In various tissues, cells are usually in contact with a complex network of the extracellular matrix (ECM) that plays an essential role in the formation, maintenance, and remodeling of tissue architecture. In fact, this ECM is composed of a large number of proteins capable of activating intracellular signaling pathways that influence a broad spectrum of biological functions, such as proliferation, adhesion, migration, and cell invasion of normal and tumor cells. The process of cellular invasion is generally controlled by the action of various proteases including the metalloproteases (MMPs). The latter induces the degradation of extracellular matrix (ECM) which promotes cell migration and invasion. Indeed, MMPs are a group of enzymes with endopeptidase activity able of degrading various components of the ECM including collagen, gelatin, proteoglycan, laminin, fibronectin, and vitronectin. Under physiological conditions, the activity of MMPs is tightly regulated [128]. Indeed, most of these proteases are not constitutively expressed by normal cells, and their expression is induced by cytokines and growth factors during wound healing or inflammation. Increased activity or expression of these MMPs was found to be involved in various pathological situations related to a considerable degradation of ECM as reported for different types of cancers such as breast, pancreatic, ovarian, and skin cancers. In addition, the level of expression and activation of these MMPs has been correlated with poor prognosis [130]. MMPs produced by tumor cells exert potent proteolytic activity, allowing ECM degrading and their invasion. In many cases, by secreting various MMPs, these cells are also able to penetrate into the stroma and colonize the targeted tissues. Under physiological conditions, the activity of MMPs such as MMP-2 and MMP-9 is regulated by their naturally occurring inhibitors, tissue inhibitors of metalloproteases (TIMPs) [131] which usually form complexes with inactive MMPs [131]. MMPs are synthesized as inactive zymogens and are activated by post-translational hydrolysis either directly by PCs as is the case for

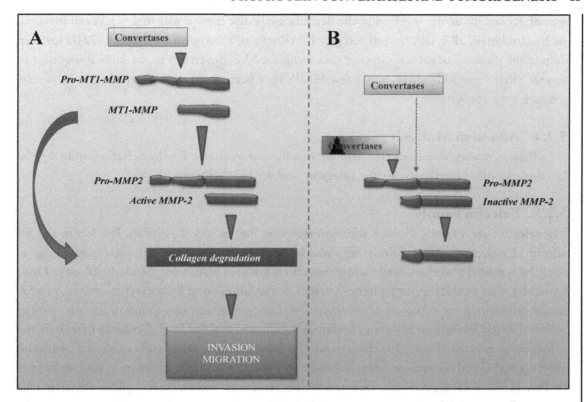

FIGURE 13: Role of PCs in the activation of pro-MT1-MMP and pro-MMP-2. Pro-MT1-MMP is activated by PCs. The active enzyme can be thus directly involved in collagen degradation that induces cell invasion and/or migration or indirectly by activating pro-MMP2 (A). Furin can also induce the inactivation of MMP2 by proteolytic cleavage at Pro-MM-2 prodomain (B).

MT-MMPs (membrane-type matrix metalloproteinases) or by other enzymes such as plasmin. In addition, activated MT-MMPs exert their extracellular enzymatic activity and induce the activation of other MMPs. Indeed, it has been shown that MMP-2 synthesized as inactive precursor is activated by MT1-MMP itself activated by PCs, suggesting the involvement of PCs in cell invasion (Figure 13A). Recently, it has been shown that in some situations still poorly known, Furin can cleave pro-MMP-2 at its prodomain and generate an inactive intermediate form of MMP-2. [132] (Figure 13B).

Using different inhibitors of the PCs, previous studies revealed that inhibition of PCs in various tumor cell lines significantly reduced their invasive potential through inhibition of MMPs processing and activity [101]. Inhibition of the convertases in tumor cells also reduced the expression of various molecules involved in the degradation of the extracellular matrix and cell invasion, such as urokinase uPA, tPA, and its receptor uPAR by a still unknown mechanism [72]. However, despite the presence of various cleavage sites recognized by convertases in the protein sequence of several MMPs and MT-MMP, only a few MMPs have been demonstrated experimentally as substrates for the convertases.

5.2.4 Adhesion Molecules

All adhesion molecules include four families of adhesion receptors: 1—the Selectin family, 2—the Immunoglobulin superfamily, 3—the Integrins, and 4—the Cadherins.

5.2.5 Selectin Family

The selectins are a family of three adhesion receptors that include L-selectin, P-selectin, and E-selectin. L-selectin is constitutively expressed by the lymphocytes, E-selectin is expressed exclusively by activated endothelial cells, and P-Selectin is found in platelets and endothelial cells. These molecules were initially reported to be involved in the adhesion of leukocytes to activated endothelial cells during the processes of inflammation. Lately, selectins were found to also play similar role during the interaction between circulating tumor cells and endothelial cells that facilitate the colonization process (Figure 14). In general, the selectins bind sugars such as sialyl-Lewis^x and sialyl-Lewis^a found on various tumor cells that play key roles in their adhesion to activated endothelium [133]. Indeed, previously, various adhesion assays revealed that colon, pancreatic, or gastric cancer cells use sialyl-Lewis^x to adhere to endothelial cells. Subsequently, the ability of cancer cells to adhere to endothelial cells was found to be directly proportional to the metastatic potential of cancer cells [134]. Indeed, cimetidine, a drug that prevents the expression of E-selectin, blocked the adhesion of tumor cells to the endothelium and inhibits metastasis [135].

During the metastasis process, the interaction between tumor cells and their environment plays a key role in the colonization process [136, 137]. Indeed, it has been shown that the arrival of metastatic tumor cells in the hepatic circulation results in the induction of the expression of various cytokines such as IL-1 and TNF-α. These molecules are responsible for the induction of the expression of different adhesion molecules including E-selectin [136]. The ability of tumor cells to induce this molecule is directly related to their metastatic capacity. It has been shown that only cells with a metastatic potential were able to induce E-selectin *in vitro* and *in vivo*. Inhibition of E-selectin by different strategies such as the use of antibodies against E-selectin or its ligands sialyl-Lewis results in decreased tumor cells adhesion to the endothelial wall and consequently a reduction in the formation of liver metastases. These studies suggest that the induction of E-selectin by tumor cells is due to the secretion of molecules able to induce E-selectin expression during their interaction with endothelial cells. Indeed, several studies have demonstrated that several colon cancer cells secrete

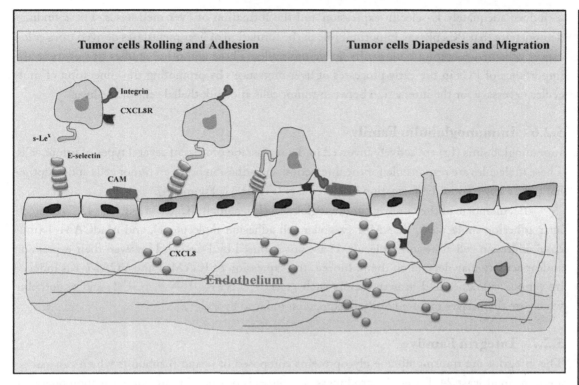

FIGURE 14: Schematic representation of tumor cells and endothelial cells interaction that involves different adhesion molecules. The adhesion of tumor cells to endothelial cells in this situation depends on the expression of selectins on endothelial cells, in particular E-selectin. This interaction will result in the rolling of tumor cells that allows the activation of integrins necessary for firm adhesion between endothelial cells and tumor cells. Thereafter, under the action of chemokines (CXCL8), tumor cells cross the endothelial cells to proliferate and form secondary metastases.

various molecules inducing the expression of E-selectin such as IL-1, IL-6, and TNF-α [137–140]. In addition, the levels of these molecules were found to be abnormally high in the supernatant of colon carcinoma cells as compared to cells derived from normal colon [140]. Also, certain growth factors produced by the tumor cells have been described as being involved in the expression of E-selectin. These include VEGF [141], IGF-1 [142], PDGF [143], and endothelin [144] and were previously reported to be activated or induced by the PCs. Previously, the expression of the general PCs inhibitor α1-PDX in tumor cells significantly reduced the ability of culture medium conditioned by these cells to induce E-selectin *in vitro*. Indeed, we previously found decreased levels of IL-1 and TNF in media derived from cells expressing α1-PDX. *In vivo*, these cells were unable to induce adequately E-selectin expression and the formation of liver metastases. These findings demonstrate that PCs play an important role in the formation of liver metastases derived from colon cancer cells by generating biologically active molecules. These and other studies demonstrate the importance of PCs in the early processes of liver metastases by promoting the generation of molecules necessary for the interaction between tumor cells and endothelial cells of the liver.

5.2.6   Immunoglobulin Family

Immunoglobulins (Ig) are actively involved in the metastatic process of several types of tumor cells. These molecules are responsible for the interaction and adhesion between tumor cells and endothelial cells, a key step in the formation of metastases (Figure 11, Figure 14).

All members of the family of immunoglobulin (Ig), namely, ICAM-1, -2, -3, -4, -5 (intercellular adhesion molecules), VCAM-1 (vascular cell adhesion molecule-1), and MadCAM-1 (mucosal addressin cell adhesion molecule-1) are not matured by PCs [2]. However, their expression and/or activity may depend on them. Indeed, the expression of ICAM-1 and VCAM-1 is induced by various cytokines and growth factors such as IL-1, TNF-α, IFN γ, IGF-1, and endothelin, which are themselves matured or their expression induced by PCs [2].

5.2.7   Integrin Family

The integrins are transmembrane glycoproteins composed of α and β subunits which can associate to form at least 24 distinct heterodimers. The ligands of integrins are common components of the ECM, such as collagen, fibronectin, vitronectin, and laminin. The interaction of cells with the extracellular matrix induces not only cytoskeleton organization but also the activation of several intracellular signaling pathways [145]. Indeed, these molecules use the extracellular domains of the two subunits to bind to their ligands, while the cytosolic parts seemed to interact with the cytoskeleton, which enables the induction of changes in motility, cell growth, and survival. Previously, it has been shown that altering the expression and/or activation of several integrins is linked to various cancers. In addition, activation of integrins is involved in MMP activity in various tumor cell types [146]. Although all the α chains of integrins contain a site recognized by the PCs, only few of them were demonstrated to be experimentally substrates of the PCs (Table 5). Previously, Berthet et al. revealed that the cleavage of αv chain by PCs is an important process for the mediation of signal transduction necessary for cell adhesion. Indeed, inhibition of the maturation of this integrin alters tumor cell adhesion to vitronectin [147]. The proteolytic inhibition of this chain was found to inhibit the phosphorylation of Focal Adhesion Kinase (FAK) and Paxillin—two molecules actively involved in cell adhesion. However, further studies are required for the clarification of the precise role of the PCs in the regulation of integrin functions. Indeed, previous studies reported that the cleavage of the integrin α4 by the PCs does not affect their interaction with their ligand and its adhesive function [148].

5.2.8 Cadherin Family

Cadherins are a family of transmembrane proteins involved in cell–cell interaction. In the presence of calcium, the extracellular portion of a cadherin interacts with the extracellular portion of another adjacent cell. Their cytoplasmic portion allows them to bind to the actin cytoskeleton via catenins [149]. During the development of epithelial cancers, such as colon cancer, stomach, liver, esophagus, skin, lung, and breast, the adhesion function of E-cadherin, which is major component of the adherent junctions type, is usually lost. This is associated with a loss of tissue cohesion and induced cell invasiveness. Thus, loss of E-cadherin may facilitate the spread of cancer cells from a tumor. Cadherins are synthesized as precursor proteins and cleaved by PCs. However, despite recent studies showing that the maturation of E-cadherin is necessary for the mediation of its function, the role of maturation of cadherins by the PCs in cancer remains unknown.

5.3   PROPROTEIN CONVERTASES IN THE CLINICAL SETTING

Although elevated expression of different PCs was reported for different human cancers and tumor cell lines, the importance of various PCs in these cancers has not yet been clarified. Early studies revealed a high Furin expression in advanced lung tumors [150]. Such an association has subsequently been confirmed in other malignancies such as breast [99], head, and neck cancers [105]. At the clinical setting, a phase I trial of FANG vaccine, an autologous tumor-based product incorporating a plasmid encoding GMCSF and a bifunctional short hairpin RNAi (bi-shRNAi) targeting Furin was recently found successful with 91% success rate in patients with advanced cancer [151]. This clinical trial revealed that FANG vaccine was safe and elicited an immune response in these patients that prolonged their survival. In contrast, recently, Furin was found to be over-expressed in liver cancer, and the high expression of Furin in HCC tissues was found to predict a better postoperative disease-free survival. The use of HCC mice experimental model revealed that over-expression of Furin inhibited HCC tumor growth in a subcutaneous xenograft model [152].

Other studies described a significant association between high expression of PC1 and PC2 in neuroendocrine tumors, suggesting their involvement in the malignancy of tumor cells with a neural and/or endocrine phenotype [153–157]. Although these studies showed a positive association between PC1 or PC2 expression and the extent of tumors, further research is required to elucidate the importance and the prognostic role of these enzymes in endocrine-related cancers.

Studies on the prognostic impact of PACE4 expression in tumors are also less conclusive.

Previously, PACE4 expression was reported to be significantly higher in breast tumors [99]. In other studies, PACE4 expression was found to be up-regulated in human head and neck tumors and tumor cell lines [150, 158]. However, this may be tissue-dependent because analysis of lung solid tumors revealed that only half expressed PACE4 and therein its mRNA level was lower than that of Furin [150].

PC5 and PC7 have been examined for their prognostic relevance in only a few human cancers. These studies showed a positive association between PC7 expression and the extent of breast tumors of which PC5 was undetectable or weakly expressed [99, 159].

. . . .

Acknowledgments

This work was supported by grants from INCA, INSERM, LLCC, Region Aquitaine, and Bordeaux-1 University.

References

[1] Seidah NG, Chrétien M. (1999) Proprotein and prohormone convertases: a family of sub-tilases generating diverse bioactive polypeptides. *Brain Research.* 848: pp. 45–62.

[2] Khatib AM, Siegfried G, Chrétien M, Metrakos P, Seidah NG. (2002) Proprotein convertases in tumor progression and malignancy: novel targets in cancer therapy. *Am J Pathol.* 160: pp. 1921–35.

[3] Varshavsky A, Kessler O, Abramovitch S, Kigel B, Zaffryar S, Akiri G, Neufeld G. (2008) Semaphorin-3B is an angiogenesis inhibitor that is inactivated by Furin-like pro-protein convertases. *Cancer Res.* 68: pp. 6922–31.

[4] Creemers JW, Ines Dominguez D, Plets E, Serneels L, Taylor NA, Multhaup G, Craessaerts K, Annaert W, De Strooper B. (2001) Processing of beta-secretase by Furin and other members of the proprotein convertase family. *J Biol Chem.* 276: pp. 4211–7.

[5] Khatib AM, Siegfried G. (2006) Growth factors: To cleave or not to cleave. In: Regulation of Carcinogenesis, Angiogenesis and Metastasis by the Proprotein Convertases: A New Potential Strategy in Cancer Therapy. Ed: Khatib AM, Springer Science. Kluwer Academic Publishers. Holland pp. 121–35.

[6] Steiner DF, Cunningham D, Spigelman L, Aten B. (1967) Insulin biosynthesis: evidence for a precursor. *Science.* 157: pp. 697–700.

[7] Chrétien M, Li CH. (1967) Isolation, purification, and characterization of gamma-lipotropic hormone from sheep pituitary glands. *Can J Biochem.* 45: pp. 1163–74.

[8] Julius D, Brake A, Blair L, Kunisawa R, Thorner J. (1984) Isolation of the putative structural gene for the lysine-arginine-cleaving endopeptidase required for processing of yeast prepro-alpha-factor. *Cell.* 37: pp. 1075–89.

[9] Fuller RS, Brake A, Thorner J. (1989) Yeast prohormone processing enzyme (KEX2 gene product) is a Ca2+–dependent serine protease. *Proc Natl Acad Sci U S A.* 86: pp. 1434–8.

[10] Roebroek AJ, Schalken JA, Leunissen JA, Onnekink C, Bloemers HP, Van de Ven WJ. (1986) Evolutionary conserved close linkage of the c-fes/fps proto-oncogene and genetic sequences encoding a receptor-like protein. *EMBO J.* 5: pp. 2197–202.

[11] Roebroek AJ, Schalken JA, Bussemakers MJ, van Heerikhuizen H, Onnekink C, Debruyne FM, Bloemers HP, Van de Ven WJ. (1986) Characterization of human c-fes/fps reveals a

new transcription unit (fur) in the immediately upstream region of the proto-oncogene. *Mol Biol Rep.* 11: pp. 117–25.

[12]   van de Ven WJ, Voorberg J, Fontijn R, Pannekoek H, van den Ouweland AM, van Duijn-hoven HL, Roebroek AJ, Siezen RJ. (1990) Furin is a subtilisin-like proprotein processing enzyme in higher eukaryotes. *Mol Biol Rep.* 14: pp. 265–75.

[13]   Seidah NG, Gaspar L, Mion P, Marcinkiewicz M, Mbikay M, Chrétien M. (1990) cDNA sequence of two distinct pituitary proteins homologous to Kex2 and Furin gene products: tissue-specific mRNAs encoding candidates for pro-hormone processing proteinases. *DNA Cell Biol.* 9: 789. Erratum for DNA Cell Biol. 9: pp. 415–24.

[14]   Seidah NG, Marcinkiewicz M, Benjannet S, Gaspar L, Beaubien G, Mattei MG, Lazure C, Mbikay M, Chrétien M. (1991) Cloning and primary sequence of a mouse candidate prohormone convertase PC1 homologous to PC2, Furin, and Kex2: distinct chromosomal localization and messenger RNA distribution in brain and pituitary compared to PC2. *Mol Endocrinol.* 5: pp. 111–22.

[15]   Smeekens SP, Steiner DF. (1990) Identification of a human insulinoma cDNA encoding a novel mammalian protein structurally related to the yeast dibasic processing protease Kex2. *J Biol Chem.* 265: pp. 2997–3000.

[16]   Kiefer MC, Tucker JE, Joh R, Landsberg KE, Saltman D, Barr PJ. (1991) Identification of a second human subtilisin-like protease gene in the fes/fps region of chromosome 15. *DNA Cell Biol.* 10: pp. 757–69.

[17]   Nakayama K, Watanabe T, Nakagawa T, Kim WS, Nagahama M, Hosaka M, Hatsuzawa K, Kondoh-Hashiba K, Murakami K. (1992b) Consensus sequence for precursor process-ing at mono-arginyl sites. Evidence for the involvement of a Kex2-like endoprotease in pre-cursor cleavages at both dibasic and mono-arginyl sites. *J Biol Chem.* 267: pp. 16335–40.

[18]   Seidah NG, Day R, Hamelin J, Gaspar A, Collard MW, Chrétien M. (1992) Testicular ex-pression of PC4 in the rat: molecular diversity of a novel germ cell-specific Kex2/subtilisin-like proprotein convertase. *Mol Endocrinol.* 6: pp. 1559–70.

[19]   Lusson J, Vieau D, Hamelin J, Day R, Chrétien M, Seidah NG. (1993) cDNA structure of the mouse and rat subtilisin/kexin-like PC5: a candidate proprotein convertase expressed in endocrine and nonendocrine cells. *Proc Natl Acad Sci U S A.* 90: pp. 6691–5.

[20]   Nakagawa T, Hosaka M, Torii S, Watanabe T, Murakami K, Nakayama K. (1993) Identifi-cation and functional expression of a new member of the mammalian Kex2-like processing endoprotease family: its striking structural similarity to PACE4. *J Biochem.* 113: pp. 132–5.

[21]   Nakagawa T, Murakami K, Nakayama K. (1993b) Identification of an isoform with an extremely large Cys-rich region of PC6, a Kex2-like processing endoprotease. *FEBS Lett.* 327: pp. 165–71.

[22]   Seidah NG, Hamelin J, Mamarbachi M, Dong W, Tardos H, Mbikay M, Chretien M,

Day R. (1996) cDNA structure, tissue distribution, and chromosomal localization of rat PC7, a novel mammalian proprotein convertase closest to yeast kexin-like proteinases. *Proc Natl Acad Sci U S A.* 93: pp. 3388–93.

[23]   Meerabux J, Yaspo ML, Roebroek AJ, Van de Ven WJ, Lister TA, Young BD. (1996) A new member of the proprotein convertase gene family (LPC) is located at a chromosome translocation breakpoint in lymphomas. *Cancer Res.* 56: pp. 448–51.

[24]   Bruzzaniti A, Goodge K, Jay P, Taviaux SA, Lam MH, Berta P, Martin TJ, Moseley JM, Gillespie MT. (1996) PC8 [corrected], a new member of the convertase family. *Biochem J.* 314: pp. 727–31.

[25]   Seidah NG, Mowla SJ, Hamelin J, Mamarbachi AM, Benjannet S, Touré BB, Basak A, Munzer JS, Marcinkiewicz J, Zhong M, Barale JC, Lazure C, Murphy RA, Chrétien M, Marcinkiewicz M. (1999b) Mammalian subtilisin/kexin isozyme SKI-1: A widely expressed proprotein convertase with a unique cleavage specificity and cellular localization. *Proc Natl Acad Sci U S A.* 96: pp. 1321–6.

[26]   Sakai J, Rawson RB, Espenshade PJ, Cheng D, Seegmiller AC, Goldstein JL, Brown MS. (1998) Molecular identification of the sterol-regulated luminal protease that cleaves SREBPs and controls lipid composition of animal cells. *Mol Cell.* 2: pp. 505–14.

[27]   Abifadel M, Varret M, Rabès JP, Allard D, Ouguerram K, Devillers M, Cruaud C, Benjannet S, Wickham L, Erlich D, Derré A, Villéger L, Farnier M, Beucler I, Bruckert E, Chambaz J, Chanu B, Lecerf JM, Luc G, Moulin P, Weissenbach J, Prat A, Krempf M, Junien C, Seidah NG, Boileau C. (2003) Mutations in PCSK9 cause autosomal dominant hypercholesterolemia. *Nat Genet.* 34: pp. 154–6.

[28]   Seidah NG, Benjannet S, Wickham L, Marcinkiewicz J, Jasmin SB, Stifani S, Basak A, Prat A, Chretien M. (2003) The secretory proprotein convertase neural apoptosis-regulated convertase 1 (NARC-1): liver regeneration and neuronal differentiation. *Proc Natl Acad Sci U S A.* 100: pp. 928–33.

[29]   Anderson ED, VanSlyke JK, Thulin CD, Jean F, Thomas G. (1997) Activation of the Furin endoprotease is a multiple-step process: requirements for acidification and internal propeptide cleavage. *EMBO J.* 16: pp. 1508–18.

[30]   Creemers JW, Vey M, Schäfer W, Ayoubi TA, Roebroek AJ, Klenk HD, Garten W, Van de Ven WJ. (1995) Endoproteolytic cleavage of its propeptide is a prerequisite for efficient transport of Furin out of the endoplasmic reticulum. *J Biol Chem.* 270: pp. 2695–702.

[31]   Anderson ED, Molloy SS, Jean F, Fei H, Shimamura S, Thomas G. (2002) The ordered and compartment-specfific autoproteolytic removal of the Furin intramolecular chaperone is required for enzyme activation. *J Biol Chem.* 277: pp. 12879–90.

[32]   Thomas G. (2002) Furin at the cutting edge: from protein traffic to embryogenesis and disease. *Nat Rev Mol Cell Biol.* 3: pp. 753–66.

[33] Siezen RJ, Leunissen JA. (1997) Subtilases: the superfamily of subtilisin-like serine proteases. *Protein Sci.* 6: pp. 501–23.

[34] Scougall K, Taylor NA, Jermany JL, Docherty K, Shennan KI. (1998) Differences in the autocatalytic cleavage of pro-PC2 and pro-PC3 can be attributed to sequences within the propeptide and Asp310 of pro-PC2. *Biochem J.* 334: pp. 531–7.

[35] Zhou A, Martin S, Lipkind G, LaMendola J, Steiner DF. (1998) Regulatory roles of the P domain of the subtilisin-like prohormone convertases. *J Biol Chem.* 273: pp. 11107–14.

[36] Lusson J, Benjannet S, Hamelin J, Savaria D, Chrétien M, Seidah NG. (1997) The integrity of the RRGDL sequence of the proprotein convertase PC1 is critical for its zymogen and C-terminal processing and for its cellular trafficking. *Biochem J.* 326: pp. 737–44.

[37] Nour N, Mayer G, Mort JS, Salvas A, Mbikay M, Morrison CJ, Overall CM, Seidah NG. (2005) The cysteine-rich domain of the secreted proprotein convertases PC5A and PACE4 functions as a cell surface anchor and interacts with tissue inhibitors of metalloproteinases. *Mol Biol Cell.* 16: pp. 5215–26.

[38] Mayer G, Hamelin J, Asselin MC, Pasquato A, Marcinkiewicz E, Tang M, Tabibzadeh S, Seidah NG. (2008) The regulated cell surface zymogen activation of the proprotein convertase PC5A directs the processing of its secretory substrates. *J Biol Chem.* 283: pp. 2373–84.

[39] Tadros H, Chrétien M, Mbikay M. (2001) The testicular germ-cell protease PC4 is also expressed in macrophage-like cells of the ovary. *J Reprod Immunol.* 49: pp. 133–52.

[40] Seidah NG, Chrétien M, Day R. (1994) The family of subtilisin/kexin like pro-protein and pro-hormone convertases: divergent or shared functions. *Biochimie.* 76: pp. 197–209.

[41] Seidah NG, Day R, Marcinkiewicz M, Chrétien M. (1998) Precursor convertases: an evolutionary ancient, cell-specific, combinatorial mechanism yielding diverse bioactive peptides and proteins. *Ann N Y Acad Sci.* 839: pp. 9–24.

[42] Essalmani R, Hamelin J, Marcinkiewicz J, Chamberland A, Mbikay M, Chrétien M, Seidah NG, Prat A. (2006) Deletion of the gene encoding proprotein convertase 5/6 causes early embryonic lethality in the mouse. *Mol Cell Biol.* 26: pp. 354–61.

[43] Molloy SS, Thomas L, VanSlyke JK, Stenberg PE, Thomas G. (1994) Intracellular trafficking and activation of the Furin proprotein convertase: localization to the TGN and recycling from the cell surface. *EMBO J.* 13: pp. 18–33.

[44] Molloy SS, Anderson ED, Jean F, Thomas G. (1999) Bi-cycling the Furin pathway: from TGN localization to pathogen activation and embryogenesis. *Trends Cell Biol.* 9: pp. 28–35.

[45] Munzer JS, Basak A, Zhong M, Mamarbachi A, Hamelin J, Savaria D, Lazure C, Hendy GN, Benjannet S, Chrétien M, Seidah NG. (1997) In vitro characterization of the novel proprotein convertase PC7. *J Biol Chem.* 272: pp. 19672–81.

[46] Schapiro FB, Soe TT, Mallet WG, Maxfield FR. (2004) Role of cytoplasmic domain serines in intracellular trafficking of Furin. *Biol Cell.* 15: pp. 2884–94.

[47]   Wan L, Molloy SS, Thomas L, Liu G, Xiang Y, Rybak SL, Thomas G. (1998) PACS-1 defines a novel gene family of cytosolic sorting proteins required for trans-Golgi network localization. *Cell.* 94: pp. 205–16.

[48]   Molloy SS, Thomas L, Kamibayashi C, Mumby MC, Thomas G. (1998) Regulation of endosome sorting by a specific PP2A isoform. *J Cell Biol.* 142: pp. 1399–411.

[49]   Gagnon J, Mayne J, Mbikay M, Woulfe J, Chrétien M. (2009) Expression of PCSK1 (PC1/3), PCSK2 (PC2) and PCSK3 (Furin) in mouse small intestine. *Regul Pept.* 152: pp. 54–60.

[50]   Marcinkiewicz M, Day R, Seidah NG, Chrétien M. (1993) Ontogeny of the prohormone convertases PC1 and PC2 in the mouse hypophysis and their colocalization with cortico-tropin and alpha-melanotropin. *Proc Natl Acad Sci U S A.* 90: pp. 4922–6.

[51]   Koide S, Yoshida I, Tsuji A, Matsuda Y. (2003) The expression of proprotein convertase PACE4 is highly regulated by Hash-2 in placenta: possible role of placenta-specific basic helix–loop–helix transcription factor, human achaete–scute homologue-2. *J Biochem.* 134: pp. 433–40.

[52]   Tsuji A, Sakurai K, Kiyokage E, Yamazaki T, Koide S, Toida K, Ishimura K, Matsuda Y. (2003) Secretory proprotein convertases PACE4 and PC6A are heparin-binding proteins which are localized in the extracellular matrix. Potential role of PACE4 in the activation of proproteins in the extracellular matrix. *Biochim Biophys Acta.* 1645: pp. 95–104.

[53]   Seidah NG, Khatib AM, Prat A. (2006) The proprotein convertases and their implication in sterol and/or lipid metabolism. *Biol Chem.* 387: pp. 871–7.

[54]   Remacle AG, Shiryaev SA, Oh ES, Cieplak P, Srinivasan A, Wei G, Liddington RC, Ratnikov BI, Parent A, Desjardins R, Day R, Smith JW, Lebl M, Strongin AY. (2008) Substrate cleavage analysis of Furin and related proprotein convertases. A comparative study. *J Biol Chem.* 283: pp. 20897–906.

[55]   Oda K, Ikeda M, Tsuji E, Sohda M, Takami N, Misumi Y, Ikehara Y. (1991) Sequence requirements for proteolytic cleavage of precursors with paired basic amino acids. *Biochem Biophys Res Commun.* 179: pp. 1181–6.

[56]   Hosaka M, Nagahama M, Kim WS, Watanabe T, Hatsuzawa K, Ikemizu J, Murakami K, Nakayama K. (1991) Arg-X-Lys/Arg-Arg motif as a signal for precursor cleavage catalyzed by Furin within the constitutive secretory pathway. *J Biol Chem.* 266: pp. 12127–30.

[57]   Watanabe T, Nakagawa T, Ikemizu J, Nagahama M, Murakami K, Nakayama K. (1992) Sequence requirements for precursor cleavage within the constitutive secretory pathway. *J Biol Chem.* 267: pp. 8270–4.

[58]   Braks JA, Martens GJ. (1994) 7B2 is a neuroendocrine chaperone that transiently interacts with prohormone convertase PC2 in the secretory pathway. *Cell.* 78: pp. 263–73.

[59]   Bartolomucci A, Possenti R, Mahata SK, Fischer-Colbrie R, Loh YP, Salton SR. (2011)

The extended granin family: structure, function, and biomedical implications. *Endocr Rev.* 32: pp. 755–97.

[60]   Lee SN, Prodhomme E, Lindberg I. (2004) Prohormone convertase 1 (PC1) processing and sorting: effect of PC1 propeptide and proSAAS. *J Endocrinol.* 182: pp. 353–64.

[61]   Cameron A, Fortenberry Y, Lindberg I. (2000) The SAAS granin exhibits structural and functional homology to 7B2 and contains a highly potent hexapeptide inhibitor of PC1. *FEBS Lett.* 473: pp. 135–8.

[62]   Apletalina E, Appel J, Lamango NS, Houghten RA, Lindberg I. (1998) Identification of inhibitors of prohormone convertases 1 and 2 using a peptide combinatorial library. *J Biol Chem.* 273: pp. 26589–95.

[63]   Zhong M, Munzer JS, Basak A, Benjannet S, Mowla SJ, Decroly E, Chrétien M, Seidah NG. (1999) The prosegments of Furin and PC7 as potent inhibitors of proprotein convertases. In vitro and ex vivo assessment of their efficacy and selectivity. *J Biol Chem.* 274: pp. 33913–20.

[64]   Nour N, Basak A, Chrétien M, Seidah NG. (2003) Structure–function analysis of the prosegment of the proprotein convertase PC5A. *J Biol Chem.* 278: pp. 2886–95.

[65]   Bhattacharjya S, Xu P, Zhong M, Chrétien M, Seidah NG, Ni F. (2000) Inhibitory activity and structural characterization of a C-terminal peptide fragment derived from the prosegment of the proprotein convertase PC7. *Biochemistry.* 39: pp. 2868–77.

[66]   Fugère M, Limperis PC, Beaulieu-Audy V, Gagnon F, Lavigne P, Klarskov K, Leduc R, Day R. (2002) Inhibitory potency and specificity of subtilase-like pro-protein convertase (SPC) prodomains. *J Biol Chem.* 277: pp. 7648–56.

[67]   Bontemps Y, Lapierre M, Siegfried G, Calvo F, Khatib AM. (2008) Inhibitory feature of the proprotein convertases prosegments. *Med Chem.* 4: pp. 116–20.

[68]   Siegfried G, Basak A, Cromlish JA, Benjannet S, Marcinkiewicz J, Chrétien M, Seidah NG, Khatib AM. (2003) The secretory proprotein convertases Furin, PC5, and PC7 activate VEGF-C to induce tumorigenesis. *J Clin Invest.* 111: pp. 1723–32.

[69]   Siegfried G, Khatib AM, Benjannet S, Chrétien M, Seidah NG. (2003) The proteolytic processing of pro-platelet-derived growth factor-A at RRKR(86) by members of the proprotein convertase family is functionally correlated to platelet-derived growth factor-A-induced functions and tumorigenicity. *Cancer Res.* 63: pp. 1458–63.

[70]   Siegfried G, Basak A, Prichett-Pejic W, Scamuffa N, Ma L, Benjannet S, Veinot JP, Calvo F, Seidah N, Khatib AM. (2005) Regulation of the stepwise proteolytic cleavage and secretion of PDGF-B by the proprotein convertases. *Oncogene.* 24: pp. 6925–35.

[71]   Lehmann M, André F, Bellan C, Remacle-Bonnet M, Garrouste F, Parat F, Lissitsky JC, Marvaldi J, Pommier G. (1998) Deficient processing and activity of type I insulin-

like growth factor receptor in the Furin-deficient LoVo-C5 cells. *Endocrinology.* 139: pp. 3763–71.

[72]   Khatib AM, Siegfried G, Prat A, Luis J, Chrétien M, Metrakos P, Seidah NG. (2001) Inhibition of proprotein convertases is associated with loss of growth and tumorigenicity of HT-29 human colon carcinoma cells: importance of insulin-like growth factor-1 (IGF-1) receptor processing in IGF-1-mediated functions. *J Biol Chem.* 276: pp. 30686–93.

[73]   López de Cicco R, Bassi DE, Zucker S, Seidah NG, Klein-Szanto AJ. (2005) Human carcinoma cell growth and invasiveness is impaired by the propeptide of the ubiquitous proprotein convertase Furin. *Cancer Res.* 65: pp. 4162–71.

[74]   Owen MC, Brennan SO, Lewis JH, Carrell RW. (1983) Mutation of antitrypsin to antithrombin. alpha 1-antitrypsin Pittsburgh (358 Met leads to Arg), a fatal bleediLOng disorder. *N Engl J Med.* 309: pp. 694–8.

[75]   Brennan SO, Owen MC, Boswell DR, Lewis JH, Carrell RW. (1984) Circulating proalbumin associated with a variant proteinase inhibitor. *Biochim Biophys Acta.* 802: pp. 24–8.

[76]   Björk I, Nordling K, Larsson I, Olson ST. (1992) Kinetic characterization of the substrate reaction between a complex of antithrombin with a synthetic reactive-bond loop tetradecapeptide and four target proteinases of the inhibitor. *J Biol Chem.* 267: pp. 19047–50.

[77]   Wright HT. (1996) The structural puzzle of how serpin serine proteinase inhibitors work. *Bioessays.* 18: pp. 453–64.

[78]   Anderson ED, Thomas L, Hayflick JS, Thomas G. (1993) Inhibition of HIV-1 gp160-dependent membrane fusion by a Furin-directed alpha 1-antitrypsin variant. *J Biol Chem.* 268: pp. 24887–91.

[79]   Decroly E, Wouters S, Di Bello C, Lazure C, Ruysschaert JM, Seidah NG. (1996) Identification of the paired basic convertases implicated in HIV gp160 processing based on in vitro assays and expression in CD4(+) cell lines. *J Biol Chem.* 271: pp. 30442–50.

[80]   Benjannet S, Savaria D, Laslop A, Munzer JS, Chrétien M, Marcinkiewicz M, Seidah NG. (1997) Alpha1-antitrypsin Portland inhibits processing of precursors mediated by proprotein convertases primarily within the constitutive secretory pathway. *J Biol Chem.* 272: pp. 26210–8.

[81]   Jean F, Stella K, Thomas L, Liu G, Xiang Y, Reason AJ, Thomas G. (1998) alpha1-Antitrypsin Portland, a bioengineered serpin highly selective for Furin: application as an antipathogenic agent. *Proc Natl Acad Sci U S A.* 95: pp. 7293–8.

[82]   Tsuji A, Kanie H, Makise H, Yuasa K, Nagahama M, Matsuda Y. (2007) Engineering of alpha1-antitrypsin variants selective for subtilisin-like proprotein convertases PACE4 and PC6: importance of the P2' residue in stable complex formation of the serpin with proprotein convertase. *Protein Eng Des Sel.* 20: pp. 163–70.

[83]   Watanabe M, Wang A, Sheng J, Gombart AF, Ayata M, Ueda S, Hirano A, Wong TC. (1995) Delayed activation of altered fusion glycoprotein in a chronic measles virus variant that causes subacute sclerosing panencephalitis. *J Neurovirol.* 1: pp. 412–23.

[84]   Scamuffa N, Calvo F, Chrétien M, Seidah NG, Khatib AM. (2006) Proprotein convertases: lessons from knockouts. *FASEB J.* 20: pp. 1954–63.

[85]   O'Rahilly S, Gray H, Humphreys PJ, Krook A, Polonsky KS, White A, Gibson S, Taylor K, Carr C. (1995) Brief report: impaired processing of prohormones associated with abnormalities of glucose homeostasis and adrenal function. *N Engl J Med.* 333: pp. 1386–90.

[86]   Jackson RS, Creemers JW, Ohagi S, Raffin-Sanson ML, Sanders L, Montague CT, Hutton JC, O'Rahilly S. (1997) Obesity and impaired prohormone processing associated with mutations in the human prohormone convertase 1 gene. *Nat Genet.* 16: pp. 303–6.

[87]   Jackson RS, Creemers JW, Farooqi IS, Raffin-Sanson ML, Varro A, Dockray GJ, Holst JJ, Brubaker PL, Corvol P, Polonsky KS, Ostrega D, Becker KL, Bertagna X, Hutton JC, White A, Dattani MT, Hussain K, Middleton SJ, Nicole TM, Milla PJ, Lindley KJ, O'Rahilly S. (2003) Small-intestinal dysfunction accompanies the complex endocrinopathy of human proprotein convertase 1 deficiency. *J Clin Invest.* 112: pp. 1550–60.

[88]   Farooqi IS, Volders K, Stanhope R, Heuschkel R, White A, Lank E, Keogh J, O'Rahilly S, Creemers JW. (2007) Hyperphagia and early-onset obesity due to a novel homozygous missense mutation in prohormone convertase 1/3. *J Clin Endocrinol Metab.* 92: pp. 3369–73.

[89]   Benzinou M, Creemers JW, Choquet H, Lobbens S, Dina C, Durand E, Guerardel A, Boutin P, Jouret B, Heude B, Balkau B, Tichet J, Marre M, Potoczna N, Horber F, Le Stunff C, Czernichow S, Sandbaek A, Lauritzen T, Borch-Johnsen K, Andersen G, Kiess W, Körner A, Kovacs P, Jacobson P, Carlsson LM, Walley AJ, Jørgensen T, Hansen T, Pedersen O, Meyre D, Froguel P. (2008) Common nonsynonymous variants in PCSK1 confer risk of obesity. *Nat Genet.* 40: pp. 943–5.

[90]   Bennett BD, Denis P, Haniu M, Teplow DB, Kahn S, Louis JC, Citron M, Vassar R. (2001) A Furin-like convertase mediates propeptide cleavage of BACE, the Alzheimer's beta-secretase. *J Biol Chem.* 275: pp. 37712–7.

[91]   Gordon VM, Klimpel KR, Arora N, Henderson MA, Leppla SH. (1995) Proteolytic activation of bacterial toxins by eukaryotic cells is performed by Furin and by additional cellular proteases. *Infect Immun.* 63: pp. 82–7.

[92]   McKee ML, FitzGerald DJ. (1999) Reduction of Furin-nicked Pseudomonas exotoxin A: an unfolding story. *Biochemistry.* 38: pp. 16507–13.

[93]   Fukui A, Horiguchi Y. (2004) Bordetella dermonecrotic toxin exerting toxicity through activation of the small GTPase Rho. *J Biochem.* 136: pp. 415–9.

[94]   Beauregard KE, Collier RJ, Swanson JA. (2000) Proteolytic activation of receptor-bound

anthrax protective antigen on macrophages promotes its internalization. *Cell Microbiol.* 2: pp. 251–8.

[95] Gordon VM, Rehemtulla A, Leppla SH. (1997) A role for PACE4 in the proteolytic activation of anthrax toxin protective antigen. *Infect Immun.* 65: pp. 3370–5.

[96] Abrami L, Fivaz M, Glauser PE, Parton RG, van der Goot FG. (1998) A pore-forming toxin interacts with a GPI-anchored protein and causes vacuolation of the endoplasmic reticulum. *J Cell Biol.* 140: pp. 525–40.

[97] Basak A, Zhong M, Munzer JS, Chrétien M, Seidah NG. (2001) Implication of the proprotein convertases Furin, PC5 and PC7 in the cleavage of surface glycoproteins of Hong Kong, Ebola and respiratory syncytial viruses: a comparative analysis with fluorogenic peptides. *Biochem J.* 353: pp. 537–45.

[98] Bergeron E, Vincent MJ, Wickham L, Hamelin J, Basak A, Nichol ST, Chrétien M, Seidah NG. (2005) Implication of proprotein convertases in the processing and spread of severe acute respiratory syndrome coronavirus. *Biochem Biophys Res Commun.* 326: pp. 554–63.

[99] Yana I, Weiss SJ. (2000) Regulation of membrane type-1 matrix metalloproteinase activation by proprotein convertases. *Mol Biol Cell.* 11: pp. 2387–401.

[100] Khatib AM. (ed). (2006) Regulation of carcinogenesis, angiogenesis and metastasis by the proprotein convertases: A new potential strategy in cancer therapy. Springer Science. Kluwer Academic Publishers. http://www.amazon.com/gp/reader/1402047932/ref=sib_dp_pt#reader-link.

[101] Bassi DE, Fu J, Lopez de Cicco R, Klein-Szanto AJ. (2005) Proprotein convertases: "master switches" in the regulation of tumor growth and progression. *Mol Carcinog.* 44: pp. 151–61. Review.

[102] Cheng M, Watson PH, Paterson JA, Seidah NG, Chrétien M, Shiu RP. (1997) Pro-protein convertase gene expression in human breast cancer. *Int J Cancer.* 71: pp. 966–71.

[103] Cheng M, Watson PH, Paterson JA, Seidah NG, Chretien M, Shiu RP. (1997) Pro-protein convertase gene expression in human breast cancer. *Int J Cancer.* 71: pp. 966–97.

[104] Bassi DE, Mahloogi H, Al-Saleem L, Lopez De Cicco R, Ridge JA, Klein-Szanto AJ. (2001) Elevated Furin expression in aggressive human head and neck tumors and tumor cell lines. *Mol Carcinog.* 31: pp. 224–32.

[105] Pardee AB. (1989) G1 events and regulation of cell proliferation. *Science.* 246: pp. 603–8.

[106] Baserga R, Rubin R. (1993) Cell cycle and growth control. *Crit Rev Eukaryot Gene Expr.* 3: pp. 47–61.

[107] Kimura I, Honda R, Okai H, Okabe M. (2000) Vascular endothelial growth factor promotes cell-cycle transition from G0 to G1 phase in subcultured endothelial cells of diabetic rat thoracic aorta. *Jpn J Pharmacol.* 83: pp. 47–55.

[108] Duguay SJ, Lai-Zhang J, Steiner DF. (1995) Mutational analysis of the insulin-like growth factor I prohormone processing site. *J Biol Chem.* 270: pp. 17566–74.

[109] Denault JB, Claing A, D'Orleans-Juste P, Sawamura T, Kido T, Masaki T, Leduc R. (1995) Processing of proendothelin-1 by human Furin convertase. *FEBS Lett.* 362: pp. 276–80.

[110] Bresnahan PA, Leduc R, Thomas L, Thorner J, Gibson HL, Brake AJ, Barr PJ, Thomas G. (1990) Human fur gene encodes a yeast KEX2-like endoprotease that cleaves pro-β-NGF in vivo. *J Cell Biol.* 111: pp. 2851–9.

[111] Seidah NG, Benjannet S, Pareek S, Savaria D, Hamelin J, Goulet B, Laliberté J, Lazure C, Chrétien M, Murphy RA. (1996) Cellular processing of the nerve growth factor precursor by the mammalian pro-protein convertases. *Biochem J.* 314: pp. 951–60.

[112] Hendy GN, Bennett HPJ, Gibbs BF, Lazure C, Day R, Seidah NG. (1995) Proparathyroid hormone is preferentially cleaved to parathyroid hormone by the prohormone convertase Furin: a mass spectrometric study. *J Biol Chem.* 270: pp. 9517–25.

[113] Dubois CM, Laprise MH, Blanchette F, Gentry LE, Leduc R. (1995) Processing of transforming growth factor beta 1 precursor by human Furin convertase. *J Biol Chem.* 270: pp. 10618–24.

[114] Liu B, Amizuka N, Goltzman D, Rabbani SA. (1995) Inhibition of processing of parathyroid hormone-related peptide by anti-sense Furin: effect in vitro and in vivo on rat Leydig (H-500) tumor cells. *Int J Cancer.* 63: pp. 276–81.

[115] Konda Y, Yokota H, Kayo T, Horiuchi T, Sugiyama N, Tanaka S, Takata K, Takeuchi T. (1997) Proprotein-processing endoprotease Furin controls the growth and differentiation of gastric surface mucous cells. *J Clin Invest.* 99: pp. 1842–51.

[116] Kayo T, Sawada Y, Suda M, Konda Y, Izumi T, Tanaka S, Shibata H, Takeuchi T. (1997) Proprotein-processing endoprotease Furin controls growth of pancreatic beta-cells. *Diabetes.* 46: pp. 1296–304.

[117] Nguyen L, Holgado-Madruga M, Maroun C, Fixman ED, Kamikura D, Fournier T, Charest A, Tremblay ML, Wong AJ, Park M. (1997) Association of the multisubstrate docking protein Gab1 with the hepatocyte growth factor receptor requires a functional Grb2 binding site involving tyrosine 1356. *J Biol Chem.* 272: pp. 20811–19.

[118] Meisenhelder J, Suh PG, Rhee SG, Hunter T. (1989) Phospholipase C-gamma is a substrate for the PDGF and EGF receptor protein-tyrosine kinases in vivo and in vitro. *Cell.* 57: pp. 1109–22.

[119] Molloy CJ, Bottaro DP, Fleming TP, Marshall MS, Gibbs JB, Aaronson SA. (1989) PDGF induction of tyrosine phosphorylation of GTPase activating protein. *Nature.* 342: pp. 711–4.

[120] Kaplan DR, Whitman M, Schaffhausen B, Pallas DC, White M, Cantley L, Roberts TM.

(1987) Common elements in growth factor stimulation and oncogenic transformation: 85 kda phosphoprotein and phosphatidylinositol kinase activity. *Cell.* 50: pp. 1021–9.

[121]   White MF, Yenush L. (1998) The IRS-signaling system: a network of docking proteins that mediate insulin and cytokine action. *Curr Top Microbiol Immunol.* 228: pp. 179–208.

[122]   Bellacosa A, Testa JR, Staal SP, Tsichlis PN. (1991) A retroviral oncogene, akt, encoding a serine-threonine kinase containing an SH2-like region. *Science.* 254: pp. 274–7.

[123]   Rodrigues GA, Park M. (1994) Oncogenic activation of tyrosine kinases. *Curr Opin Genet Dev.* 4: pp. 15–24.

[124]   Hiscox SE, Hallett MB, Puntis MC, Nakamura T, Jiang WG. (1997) Expression of the HGF/SF receptor, c-met, and its ligand in human colorectal cancers. *Cancer Invest.* 15: pp. 513–21.

[125]   Komada M, Hatsuzawa K, Shibamoto S, Ito F, Nakayama K, Kitamura N. (1993) Proteolytic processing of the hepatocyte growth factor/scatter factor receptor by Furin. *FEBS Lett.* 328: pp. 25–9.

[126]   Robertson BJ, Moehring JM, Moehring TJ. (1993) Defective processing of the insulin receptor in an endoprotease-deficient Chinese hamster cell strain is corrected by expression of mouse Furin. *J Biol Chem.* 268: pp. 24274–7.

[127]   Hwang JB, Hernandez J, Leduc R, Frost SC. (2000) Alternative glycosylation of the insulin receptor prevents oligomerization and acquisition of insulin-dependent tyrosine kinase activity. *Biochim Biophys Acta.* 1499: pp. 74–84.

[128]   Fukumoto S. (2005) Post-translational modification of Fibroblast Growth Factor 23. *Ther Apher Dial.* 9: pp. 319–22.

[129]   Chakraborti S, Mandal M, Das S, Mandal A, Chakraborti T. (2003) Regulation of matrix metalloproteinases: an overview. *Mol Cell Biochem.* 253: pp. 269–85.

[130]   Basset P, Okada A, Chenard MP, Kannan R, Stoll I, Anglard P, Bellocq JP, Rio MC. (1997) Matrix metalloproteinases as stromal effectors of human carcinoma progression: therapeutic implications. *Matrix Biol.* 15: pp. 535–41.

[131]   Baker AH, Edwards DR, Murphy G. (2002) Metalloproteinase inhibitors: biological actions and therapeutic opportunities. *J Cell Sci.* 115: pp. 3719–27.

[132]   Cao J, Rehemtulla A, Pavlaki M, Kozarekar P, Chiarelli C. (2005) Furin directly cleaves proMMP-2 in the trans-Golgi network resulting in a nonfunctioning proteinase. *J Biol Chem.* 280: pp. 10974–80.

[133]   Takada A, Ohmori K, Yoneda T, Tsuyuoka K, Hasegawa A, Kiso M, Kannagi R. (1993) Contribution of carbohydrate antigens sialyl Lewis A and sialyl Lewis X to adhesion of human cancer cells to vascular endothelium. *Cancer Res.* 53: pp. 354–61.

[134]   Sawada R, Tsuboi S, and Fukuda M. (1994) Differential E-selectin-dependent adhesion

efficiency in sublines of a human colon cancer exhibiting distinct metastatic potentials. *J Biol Chem.* 269: pp. 1425–31.

[135]   Kobayashi K, Matsumoto S, Morishima T, Kawabe T, Okamoto T. (2000) Cimetidine inhibits cancer cell adhesion to endothelial cells and prevents metastasis by blocking E-selectin expression. *Cancer Res.* 60: pp. 3978–84.

[136]   Khatib AM, Kontogiannea M, Fallavollita L, Jamison B, Meterissian S, Brodt P. (1999) Rapid induction of cytokine and E-selectin expression in the liver in response to metastatic tumor cells. *Cancer Res.* 59: pp. 1356–61.

[137]   Mantovani A, Bussolino F, Introna M. (1997) Cytokine regulation of endothelial cell function: from molecular level to the bedside. *Immunol Today.* 18: pp. 231–40.

[138]   Simiantonaki N, Jayasinghe C, Kirkpatrick, CJ. (2002) Effect of pro-inflammatory stimuli on tumor cell-mediated induction of endothelial cell adhesion molecules in vitro. *Exp Mol Pathol.* 73: pp. 46–53.

[139]   Takeda K, Fujii N, Nitta Y, Sakihara H, Nakayama K, Rikiishi H, Kumagai K. (1991) Murine tumor cells metastasizing selectively in the liver: ability to produce hepatocyte-activating cytokines interleukin-1 and/or -6. *Jpn J Cancer Res.* 82: pp. 1299–308.

[140]   Salman H, Bergman M, Bessler H, Wolloch Y, Punsky I, Djaldetti M. (2000) Effect of colon carcinoma cell supernatants on cytokine production and phagocytic capacity. *Cancer Lett.* 159: pp. 197–203.

[141]   Kim I, Moon SO, Kim SH, Kim HJ, Koh YS, Koh GY. (2001) VEGF expression of ICAM-1, VCAM-1 and E-selectin through nuclear factor-B activation in endothelial cells. *J Biol Chem.* 276: pp. 7614–20.

[142]   Che W, Lerner-Marmarosh N, Huang Q, Osawa M, Ohta S, Yoshizumi M, Glassman M, Lee JD, Yan C, Berk BC, Abe J. (2002) Insulin-like growth factor-1 enhances inflammatory responses in endothelial cells: role of Gab1 and MEKK3 in TNF-alpha-induced c-Jun and NF-kappaB activation and adhesion molecule expression. *Circ Res.* 90: pp. 1222–30.

[143]   Shankar R, de la Motte CA, Poptic EJ, DiCorleto PE. (1994) Thrombin receptor-activating peptides differentially stimulate platelet-derived growth factor production, monocytic cell adhesion, and E-selectin expression in human umbilical vein endothelial cells. *J Biol Chem.* 269: pp. 13936–41.

[144]   McCarron RM, Wang L, Stanimirovic DB, Spatz M. (1993) Endothelin induction of adhesion molecule expression on human brain microvascular endothelial cells. *Neurosci Lett.* 156: pp. 31–4.

[145]   Rigot V, and Luis J. (2006) Modulation of integrin function by endoproteolytic processing: role in tumor progression. In: Regulation of Carcinogenesis, Angiogenesis and Metastasis

by the Proprotein Convertases: A New Potential Strategy in Cancer Therapy. Ed: Khatib AM, Springer Science. Kluwer Academic Publishers. Holland, pp. 107–19.

[146]   Das S, Banerji A, Frei E, Chatterjee A. (2008) Rapid expression and activation of MMP-2 and MMP-9 upon exposure of human breast cancer cells (MCF-7) to fibronectin in serum free medium. *Life Sci.* 82: pp. 467–76.

[147]   Berthet V, Rigot V, Champion S, Secchi J, Fouchier F, Marvaldi J, Luis J. (2000) Role of endoproteolytic processing in the adhesive and signaling functions of alphavbeta5 integrin. *J Biol Chem.* 275: pp. 33308–13.

[148]   Eric Bergeron, Ajoy Basak, Etienne Decroly, Nabil G Seidah. (2003) Processing of alpha4 integrin by the proprotein convertases: histidine at position P6 regulates cleavage. *Biochem J.* 373: pp. 475–84.

[149]   Semb H, Christofori G. (1998) The tumor-suppressor function of E-cadherin. *Am J Hum Genet.* 63: pp. 1588–93.

[150]   Mbikay M, Sirois F, Yao J, Seidah NG, Chretien M. (1997) Comparative analysis of expression of the proprotein convertases Furin, PACE4, PC1 and PC2 in human lung tumours. *Br J Cancer.* 75: pp. 1509–14.

[151]   Senzer N, Barve M, Kuhn J, Melnyk A, Beitsch P, Lazar M, Lifshitz S, Magee M, Oh J, Mill SW, et al. (2012) Phase I trial of "bi-shRNAi(Furin)/GMCSF DNA/autologous tumor cell" vaccine (FANG) in advanced cancer. *Mol Ther.* 20: pp. 679–86.

[152]   Huang YH, Lin KH, Liao CH, Lai MW, Tseng YH, Yeh CT. (2012) Furin overexpression suppresses tumor growth and predicts a better postoperative disease-free survival in hepatocellular carcinoma. *PLoS One.* 7: e40738. doi: 10.1371/journal.pone.0040738. Epub 2012 Jul 10.

[153]   Takumi I, Steiner DF, Sanno N, Teramoto A, Osamura RY. (1998) Localization of prohormone convertases 1/3 and 2 in the human pituitary gland and pituitary adenomas: analysis by immunohistochemistry, immunoelectron microscopy, and laser scanning microscopy. *Mod Pathol.* 11: pp. 232–8.

[154]   Jin L, Kulig E, Qian X, Scheithauer BW, Young WF, Jr, Davis DH, Seidah NG, Chretien M, Lloyd RV. (1999) Distribution and regulation of proconvertases PC1 and PC2 in human pituitary adenomas. *Pituitary.* 1: pp. 187–95.

[155]   Kajiwara H, Itoh Y, Itoh J, Yasuda M, Osamura RY. (1999) Immunohistochemical expressions of prohormone convertase (PC)1/3 and PC2 in carcinoids of various organs. *Tokai J Exp Clin Med.* 24: pp. 13–20.

[156]   Rovere C, Barbero P, Maoret JJ, Laburthe M, Kitabgi P. (1998) Pro-neurotensin/neuromedin N expression and processing in human colon cancer cell lines. *Biochem Biophys Res Commun.* 246: pp. 155–9.

[157]   Tzimas GN, Chevet E, Jenna S, Nguyên DT, Khatib AM, Marcus V, Zhang Y, Chrétien M, Seidah N, Metrakos P. (2005) Abnormal expression and processing of the proprotein convertases PC1 and PC2 in human colorectal liver metastases. *BMC Cancer.* 5: pp. 149.

[158]   Bassi DE, Fu J, Lopez de Cicco R, Klein-Szanto AJ. (2005) Proprotein convertases: "master switches" in the regulation of tumor growth and progression. *Mol Carcinog.* 3: pp. 151–61.

[159]   Sun X, Essalmani R, Seidah NG, Prat A. (2009) The proprotein convertase PC5/6 is protective against intestinal tumorigenesis: in vivo mouse model. *Mol Cancer.* 8: p. 73.

Author Biography

Dr. Majid Khatib obtained his PhD from the University of Paris in 1997 and then completed his postdoctoral training at McGill University (Montreal, Canada). Starting in 2002, he directed his own laboratory as Scientist at the Ottawa Hospital Research Institute (OHRI), University of Ottawa (Ontario, Canada). Since 2008, he has been a research director at INSERM, first at the Hospital St-Louis in Paris and now at the University of Bordeaux. While a postdoctoral fellow, he demonstrated experimentally the implications of the proprotein convertases in the malignant phenotype of tumor cells. His work revealed at that time that the inhibition of the proprotein convertases may constitute a new potential opportunity in anticancer therapy. His research team has since discovered and characterized various substrates of these proteases and demonstrated the importance of their processing in the mediation of neoplasia. The laboratory of Dr. Khatib is also interested in the identification of small molecules inhibitors able to block the maturation of the PC substrates and reduce the malignant phenotype of tumor cells. Dr. Khatib has published many articles and reviews dealing with the importance of the proprotein convertases and their inhibitors in cancer.

TITLES OF RELATED INTEREST

Colloquium Series on Protein Activation and Cancer

Published Titles

Therapeutic Potential of Furin Inhibition: An Evaluation Using a Conditional Furin Knockout Mouse Model
Jeroen Declercq, Prof. Dr. J.W.M. Creemers
November 2012

Non-peptide Inhibitors of Proprotein Convertase Subtilisin Kexins (PCSKs): An Overall Review of Existing and New Data
Utpal Chandra De, Priyambada Mishra, Prasenjit Rudra Pal,
Biswanath Dinda, Ajoy Basak
September 2012

Proprotein Convertases in Gynecological Cancers
Andres J.P. Klein-Szanto, Jirong Zhang, Daniel Bassi
August 2012

The Role of Proprotein Convertases in Animal Models of Skin Carcinogenesis
Daniel Bassi, Jian Fu, Jirong Zhang, Andres J.P. Klein-Szanto
July 2012

Forthcoming Titles

Furin and Inflammation
Martine Cohen-Solal, *INSERM*

Processing of VEGFc and VEGFd by the PCs and Tumorigenesis
Geraldine Siegfried, *INSERM*

The Role of Proprotein Convertases in Cancer Progression and Metastasis
Peter Metrakos, *McGill University*

For a full list of published and forthcoming titles:
http://www.morganclaypool.com/page/pac/1/1

SERIES OF RELATED INTEREST

Colloquium Series on
The Building Blocks of the Cell:
Cell Structure and Function

Editor

Ivan Robert Nabi, *Professor, University of British Columbia, Department of Cellular and Physiological Sciences*

This Series is a comprehensive, in-depth review of the key elements of cell biology including 14 different categories, such as Organelles, Signaling, and Adhesion. All important elements and interactions of the cell will be covered, giving the reader a comprehensive, accessible, authoritative overview of cell biology. All authors are internationally renowned experts in their area.

For a full list of published and forthcoming titles:
http://www.morganclaypool.com/page/bbc

Colloquium Series on
The Cell Biology of Medicine

Editors

Philip L. Leopold, PhD, *Professor and Director, Department of Chemistry, Chemical Biology, & Biomedical Engineering, Stevens Institute of Technology*

Joel Pardee, Ph.D. *President, Neural Essence; formerly Associate Professor and Dean of Graduate Research, Weill Cornell School of Medicine*

In order to learn we must be able to remember, and in the world of science and medicine we remember what we envision, not what we hear. It is with this essential precept in mind that we offer the Cell Biology of Medicine series. Each book is written by faculty accomplished in teaching the scientific basis of disease to both graduate and medical students. In this modern age it has become abundantly clear that everyone is vastly interested in how our bodies work and what has gone wrong in disease. It is likewise evident that the only way to understand medicine is to engrave in our mind's eye a clear vision of the biological processes that give us the gift of life. In these lectures, we are dedicated to holding up for the viewer an insight into the biology behind the body. Each lecture demonstrates cell, tissue and organ function in health and disease. And it does so in a visually striking style. Left to its own devices, the mind will quite naturally remember the pictures. Enjoy the show.

For a list of published and forthcoming titles:
http://www.morganclaypool.com/toc/cbm/1/1

Colloquium Series on Developmental Biology

Editors

Jean-Pierre Saint-Jeannet, Ph.D., *Professor, Department of Basic Science & Craniofacial Biology, College of Dentistry, New York University*

Daniel S. Kessler, Ph.D., *Associate Professor of Cell and Developmental Biology, Chair, Developmental, Stem Cell and Regenerative Biology Program of CAMB, University of Pennsylvania School of Medicine*

Developmental biology is in a period of extraordinary discovery and research. This field will have a broad impact on the biomedical sciences in the coming decades. Developmental Biology is interdisciplinary and involves the application of techniques and concepts from genetics, molecular biology, biochemistry, cell biology, and embryology to attack and understand complex developmental mechanisms in plants and animals, from fertilization to aging. Many of the same genes that regulate developmental processes underlie human regulatory gene disorders such as cancer and serve as the genetic basis of common human birth defects. An understanding of fundamental mechanisms of development is providing a basis for the design of gene and cellular therapies for the treatment of many human diseases. Of particular interest is the identification and study of stem cell populations, both natural and induced, which is opening new avenues of research in development, disease, and regenerative medicine. This eBook series is dedicated to providing mechanistic and conceptual insight into the broad field of Developmental Biology. Each eBook is intended to be of value to students, scientists and clinicians in the biomedical sciences.

For a full list of published and forthcoming titles:
http://www.morganclaypool.com/toc/deb/1/1

Colloquium Series on
The Developing Brain

Editor

Margaret McCarthy, PhD., *Professor and Chair, Department of Pharmacology, University of Maryland School of Medicine*

The goal of this series is to provide a comprehensive state-of-the art overview of how the brain develops and those processes that affect it. Topics range from the fundamentals of axonal guidance and synaptogenesis prenatally to the influence of hormones, sex, stress, maternal care and injury during the early postnatal period to an additional critical period at puberty. Easily accessible expert reviews combine analyses of detailed cellular mechanisms with interpretations of significance and broader impact of the topic area on the field of neuroscience and the understanding of brain and behavior.

For a list of published and forthcoming titles:
http://www.morganclaypool.com/toc/dbr/1/1

Colloquium Series on
The Genetic Basis of Human Disease

Editor

Michael Dean, Ph.D., *Head, Human Genetics Section, Senior Investigator, Laboratory of Experimental Immunology National Cancer Institute (at Frederick)*

This series will explore the genetic basis of human disease, documenting the molecular basis for rare and common Mendelian and complex conditions. The series will overview the fundamental principles in understanding such as Mendel's laws of inheritance, and genetic mapping through modern examples. In addition current methods (GWAS, genome sequencing) and hot topics (epigenetics, imprinting) will be introduced through examples of specific diseases.

For a full list of published and forthcoming titles:
http://www.morganclaypool.com/page/gbhd

Colloquium Series on
Genomic and Molecular Medicine

Editor

Professor Dhavendra Kumar, MD, FRCP, FRCPCH, FACMG, *Consultant in Clinical Genetics, All Wales Medical Genetics Service Genomic Policy Unit, The University of Glamorgan, UK Institute of Medical Genetics, Cardiff University School of Medicine, University Hospital of Wales*

From 1970 onwards, there has been a continuous and growing recognition of the molecular basis of medical practice. Alongside the developments and progress in molecular medicine, new and rapid discoveries in genetics have led to an entirely new approach to the practice of clinical medicine. Until recently the field of genetic medicine has largely been restricted to the diagnosis of disease, offering explanation and assistance to patients and clinicians in dealing with a number of relatively uncommon inherited disorders. However, since the completion of the human genome in 2003 and several other genomes, there is now a plethora of information available that has attracted the attention of molecular biologists and allied researchers. A new biological science of Genomics is now with us, with far reaching dimensions and applications.

During the last decade, rapid progress has been made in new genome-level diagnostic and prognostic laboratory methods, and revealing findings in genomics have led to changes in our understanding of fundamental concepts in cell and molecular biology. It may well be that evolutionary and morbid changes at the genome level could be the basis of normal human variation and disease. Applications of individual genomic information in clinical medicine have led to the prospect of robust evidence-based personalized medicine, and genomics has led to the discovery and development of a number of new drugs with far reaching implications in pharmacotherapeutics. The existence of Genomic Medicine around us is inseparable from molecular medicine, and it contains tremendous implications for the future of clinical medicine.

For a full list of published and forthcoming titles:
http://www.morganclaypool.com/toc/gmm/1/1

Colloquium Series on Integrated Systems Physiology: From Molecule to Function to Disease

Editors

D. Neil Granger, Ph.D., *Boyd Professor and Head of the Department of Molecular and Cellular Physiology at the LSU Health Sciences Center, Shreveport*

Joey P. Granger, Ph.D., *Billy S. Guyton Distinguished Professor, Professor of Physiology and Medicine, Director of the Center for Excellence in Cardiovascular-Renal Research, and Dean of the School of Graduate Studies in the Health Sciences at the University of Mississippi Medical Center*

Physiology is a scientific discipline devoted to understanding the functions of the body. It addresses function at multiple levels, including molecular, cellular, organ, and system. An appreciation of the processes that occur at each level is necessary to understand function in health and the dysfunction associated with disease. Homeostasis and integration are fundamental principles of physiology that account for the relative constancy of organ processes and bodily function even in the face of substantial environmental changes. This constancy results from integrative, cooperative interactions of chemical and electrical signaling processes within and between cells, organs and systems. This eBook series on the broad field of physiology covers the major organ systems from an integrative perspective that addresses the molecular and cellular processes that contribute to homeostasis. Material on pathophysiology is also included throughout the eBooks. The state-of the art treatises were produced by leading experts in the field of physiology. Each eBook includes stand-alone information and is intended to be of value to students, scientists, and clinicians in the biomedical sciences. Since physiological concepts are an ever-changing work-in-progress, each contributor will have the opportunity to make periodic updates of the covered material.

For a full list of published and forthcoming titles:
http://www.morganclaypool.com/toc/isp/1/1

Colloquium Series on Neuropeptides

Editors

Lakshmi Devi, Ph.D., *Professor, Department of Pharmacology and Systems Therapeutics, Associate Dean for Academic Enhancement and Mentoring, Mount Sinai School of Medicine, New York*

Lloyd D. Fricker, Ph.D., *Professor, Department of Molecular Pharmacology, Department of Neuroscience, Albert Einstein College of Medicine, New York*

Communication between cells is essential in all multicellular organisms, and even in many unicellular organisms. A variety of molecules are used for cell-cell signaling, including small molecules, proteins, and peptides. The term 'neuropeptide' refers specifically to peptides that function as neurotransmitters, and includes some peptides that also function in the endocrine system as peptide hormones. Neuropeptides represent the largest group of neurotransmitters, with hundreds of biologically active peptides and dozens of neuropeptide receptors known in mammalian systems, and many more peptides and receptors identified in invertebrate systems. In addition, a large number of peptides have been identified but not yet characterized in terms of function. The known functions of neuropeptides include a variety of physiological and behavioral processes such as feeding and body weight regulation, reproduction, anxiety, depression, pain, reward pathways, social behavior, and memory. This series will present the various neuropeptide systems and other aspects of neuropeptides (such as peptide biosynthesis), with individual volumes contributed by experts in the field.

For a list of published and forthcoming titles:
http://www.morganclaypool.com/toc/npe/1/1

Colloquium Series on
Stem Cell Biology

Editor

Wenbin Deng, Ph.D., *Cell Biology and Human Anatomy, Institute for Pediatric Regenerative Medicine, School of Medicine, University of California, Davis*

This Series is interested in covering the fundamental mechanisms of stem cell pluripotency and differentiation, as well as strategies for translating fundamental developmental insights into discovery of new therapies. The emphasis is on the roles and potential advantages of stem cells in developing, sustaining, and restoring tissue after injury or disease. Some of the topics covered include the signaling mechanisms of development and disease; the fundamentals of stem cell growth and differentiation; the utilities of adult (somatic) stem cells, induced pluripotent stem (iPS) cells and human embryonic stem (ES) cells for disease modeling and drug discovery; and finally the prospects for applying the unique aspects of stem cells for regenerative medicine. We hope this Series will provide the most accessible and current discussions of the key points and concepts in the field, and that students and researchers all over the world will find these in-depth reviews to be useful.

For a list of published and forthcoming titles:
http://www.morganclaypool.com/page/scb

SERIES UNDER DEVELOPMENT

Colloquium Series on Drug Development

Editor

László Kürti, Ph.D., *Assistant Professor, Department of Biochemistry, University of Texas, Southwestern Medical Center, Dallas*

The Colloquium Series on Drug Development will consist of 65- to 100-page eBooks that discuss the history of a particular disease, with an emphasis on the milestones achieved in regard to the mapping-out of the underlying biology and biochemistry of that disease. 13 different categories of disease will be covered (listed below). The molecular targets (e.g. enzymes, receptors) of individual diseases will be listed in tabular format along with the available medicines. They will conclude with a look at future drug developments by reviewing basic research and clinical studies currently underway.

Colloquium Series on
Systems Biology and Data Integration

Editors

Aristotelis Tsirigos, Ph.D., *Research Scientist, IBM Computational Biology Center, IBM Research*
Gustavo Stolovitzky, Ph.D., *Manager, Functional Genomics and Systems Biology, IBM Computational Biology Center, IBM Research*

Since the beginning of the 21st century, the development of high-throughput techniques has accelerated discovery in biology. The influx of an unprecedented amount of data presents challenges as well as great opportunities for improving our understanding of living systems. Systems Biology is an interdisciplinary field which, by integrating diverse types of data, such as Genomics, Epigenomics, Proteomics, Metabolomics, etc., aims at modeling biological processes at a systems level - tissue, organ, organism - both in normal function and under stress. This eBook series is dedicated to an in-depth presentation of topics in Systems Biology of research concerning human disease. Each book is written by an expert research scientist who has extensive experience in a particular model system of disease and has demonstrated in his work the value of integrating multiple types of data with potential practical therapeutic applications. Its intended audience is students and scientists in the biomedical sciences who are interested in participating in this fascinating research game of discovery.

For a full list of published and forthcoming titles:
http://www.morganclaypool.com/page/sbdi

Colloquium Series on Molecular Mechanisms in Critical Care Medicine

Editor

Lew Romer, M.D., *Associate Professor of Anesthesiology and Critical Care Medicine, Cell Biology, Biomedical Engineering, and Pediatrics, Center for Cell Dynamics, Johns Hopkins University School of Medicine*

The idea is to provide researchers and critical care fellows with the context and tools to address the burgeoning data stream on the genetic and molecular basis of critical illness. This will be clustered around classical systems including CNS, Respiratory, Cardiovascular, Hepatic, Renal, and Hematologic.

We will explore major themes and categories of current research and practice, such as: to identify genes and molecules involved in control mechanisms for homeostasis and injury response; to identify developmental processes in which these genetic and molecular control mechanisms have a role; and to identify diseases in which these genetic and molecular control mechanisms go awry. The Series will also seek to explain and anticipate mechanisms by which critical care interventions modulate or exacerbate dysfunction of these genes and molecules. Similarly, this Series will seek to describe the role of genotypic variation in explaining the variable phenotypes of critical illness and the interaction between genetic and environmental factors in producing the variable phenotype of critical illness.

For a full list of published and forthcoming titles:
http://www.morganclaypool.com/page/mmccm

Colloquium Series on Neuroglia From Physiology to Disease

Editors

Alexej Verkhratsky, Ph.D., Professor of Neurophysiology, University of Manchester
Vlad Parpura, Ph.D., Associate Professor of Neurophysiology, University of Alabama at Birmingham

For decades the neuroglia were known, and known to be numerous, in the neural system. Nevertheless they were long thought to play only a minor, supporting role to other cells such as axons and neurons. Glial cell are now recognized are essential to neural functioning and represent an exciting, rapidly growing field in the neurosciences. This series will explore the overall molecular physiology of glial cells as well as their role in pathologic conditions.

For a full list of published and forthcoming titles:
http://www.morganclaypool.com/page/neuroglia